A SHIP BOUND TO THE FUTURE

Horizon in Social Value Creation
Where Local Insiders Meets Urban Outsiders

For Publication

Shigeru Sugawara, Mayor of Kesennuma City

Congratulations on publishing *A SHIP BOUND TO THE FUTURE— Horizon in Social Value Creation Where Local Insiders Meets Urban Outsiders.*

Kesennuma City suffered unprecedented damage from the Great East Japan Earthquake. Shipyards in the Namiita area were also devastated. The ground sank as much as 70 centimeters due to the disaster. It seemed almost impossible to resume operations at the shipyard.

People at shipbuilding which is one of the core industries, however, restored their damaged facilities in surprisingly short time with an indomitable spirit, and were relentlessly engaged in repair work of damaged fishing boats. Gradually, they began to build new ships to replace the lost ones. What we witnessed was their great hidden potentiality in supporting the Japanese fishing industry and their sincere attitude in serving their customers.

People at shipyard remained overloaded, however, by unresolved tough issues such as aging facilities, limited capability of handing larger fishing boats, aging workforce, and fundamental anxiety about the future of boat fishery.

Under this tough circumstance, or rather taking this situation as an opportunity, they carefully looked to their future, and made a very determined and astonishing decision. The four rival shipbuilding companies were merged and built their new shipyard with related industries involved. The new shipyards adopted a so called shiplift system which had only two precedents in Japan by then. The new site and docks were surrounded by unthinkable size of tide embankments in other shipyard, which reflected critical lesson learned from the experience of the tsunami. This shipyard was named MIRAI SHIPS INC. Lots of hardship by the quake ignited fire in everybody's mind and unleashed this dynamic movement, which brought about hope and courage not only me but to everyone else.

The people behind this historic project were the Shipbuilding and Ship

Machinery Division, Maritime Bureau, the Ministry of Land, Infrastructure, Transport and Tourism, Ministry of Land, Infrastructure, Transport and Tourism. They quickly set up a subsidy program for damaged shipyards and sent a person-in-charge to our city. We would like to express my sincere gratitude to the successive Directors, Deputy Directors, Section Chiefs and all those involved.

I am also grateful to the members of the Nippon Foundation for serving as the contact point for the new subsidy system and for providing various kind of support.

And, let me express my special appreciation to the members at Deloitte Tohmatsu. They showed us a very professional performance in indispensable fields and led the project to be materialized. I believe that the project did not happen without their support in terms of their proposal towards a merger by four damaged shipbuilding in the quake, their advisory in making business plan of the new company and their communication support in getting large bank loans for finance. Above all, it was highly noteworthy that, with enthusiastic mind, they kept running with top management of each company and brought them to this outcome. I would very much like them to keep escorting the new company in both strict and warm way.

The completion of MIRAI SHIPS INC. project is a symbol of recovery, and at the same time delivery of a leading entity in the evolution of fishery clusters which our city is aiming for. I have a very high hope that the MIRAI SHIPS INC. as a nationwide brand will extend its business field from the current construction and repair of fishing boats, and will make great progress as a symbol of hope to the affected areas.

I appreciate that our city was able to take advantage of the government's revival subsidy project called the Tsunami Reconstruction Base Banking Project, which allowed us to secure the site for this project. What I leaned most from this project was that, "It is the people who get things done. When the people with high aspiration get united, enormous power will be created."

Finally, as my message for publication, let me express my sincere respect and gratitude to everyone involved in the birth of MIRAI SHIPS INC.

Contents

1 | THE HISTORICAL BACKGROUND OF MIRAI SHIPS INC.

2 | THE SEVEN BUSINESS OWNERS

3 | From the Frontline of Solving Social Issues

Afterword

1 | THE HISTORICAL BACKGROUND OF MIRAI SHIPS INC.

People Engaged in Shipbuilding

It is said that there are only about ten shipyards in Japan capable of building fishing boats of 100 tons or more. In April 2008, the number of these shipyards was reduced by one. This is because a new company, MIRAI SHIPS INC., was formed through a merger.

The MIRAI SHIPS INC. was formed by four shipbuilding companies based in Kesennuma. Of these four companies, two companies, Kidoura Shipyard and Yoshida Shipyard, which built and repaired fishing boats[1], originally belonged to the above-mentioned ten companies. These two companies came together with Kosaba Shipyard and Sawada Shipyard, shipbuilding companies specializing in repairing of fishing boats[2], and together these four companies merged to form MIRAI SHIPS INC.

As Kesennuma is the largest fishing port in Japan for bonito catches and the largest or second largest fishing port in Honshu for saury catches, the fish processing industry as well as the shipbuilding industry has flourished in this area. But as everyone knows, this area was hit by the Great East Japan Earthquake on March 11, 2011, and was severely damaged. Since that day, many have gone out of business and have left their loved place. The central figures of this book are those people who chose to remain here and work for recovery and revival. But their path was never smooth.

Kidoura Shipyard and Yoshida Shipyard were once fierce rivals. But given that the shipbuilding industry has been suffering from a severely deteriorating business environment for many years, and has been impacted by trends

---

1   The representative of Kidoura Shipyard was Mr. Takeyoshi Kidoura (current President of MIRAI SHIPS INC.). The representative of Yoshida Shipyard was Mr. Keigo Yoshida (current Chairman of MIRAI SHIPS INC.). Kidoura Shipyard was founded in 1932, and Yoshida Shipyard was founded in 1916.
   Both companies built and repaired steel boats. They had almost the same number of employees (over 40 people), and were key players in the shipbuilding industry in Kesennuma.
2   The representative of Kosaba Shipyard was Mr. Yoshinori Yoshida, and the representative of Sawada Shipyard was Mr. Tsutomu Sawada. They are currently full-time directors of MIRAI SHIPS INC. Kosaba Shipyard was founded in 1910, and Sawada Shipyard was founded in 1954. Both companies repaired steel boats.

in the fishing industry which is facing a decline in catches, the two companies explored their survival by a kind of gentlemen's agreement, such as differentiating ships to build by fish types, and not encroaching upon respective customer base. Yet, in spite of the shipbuilding companies in Kesennuma being clustered within the fishing port area, and in spite of these two companies being located side by side, they were not in a collaborative relationship. The situation was such that not only top management level but even the employees barely spoke to each other. This situation, however, changed completely in the wake of the Great

Slipway system. The photo was shot right before the ship was lifted (out of the sea) and mounted onto the berth in the front for repairing.

East Japan Earthquake, and there was no other way but to join hands. When you repair a ship, you have to lift it from the sea onto the land first. The repair work starts after pulling the ship up onto an inclined "building berth" with a hoisting device. This is called the "slipway system."

A new ship is also built on a building berth. One shipyard has approximately three to six building berths. Meanwhile, as there is dense traffic of fishing boats entering the port with bonito in summer and with saury in fall, it is necessary to manage building berths during busy seasons. Thus, the companies basically had a relationship of sharing the work with each other, for example, between Kidoura Shipyard and Kosaba Shipyard or between Yoshida Shipyard and Sawada Shipyard, or between other companies. There was a common understanding among the companies that the shipbuilding industry cannot exist without the fish processing industry, and they all worked together to manage their busy seasons. This relationship laid the ground for the merger.

In addition to these four companies, there were three more companies that played an important role. Two of them are Onodera Iron Works, specializing in marine machinery and equipment, and Ishikawadenso, specializing in electrical equipment[3]. Both companies had access to shipyard sites of each shipbuilding company and had close relations with all of the four shipbuilding companies. A characteristic feature of the Kesennuma shipbuilding industry is horizontal collaboration between shipbuilding companies and such specialized companies. In general, shipbuilding companies do business directly to each specialized company, and there is no vertical relationship between them such as contractors versus sub-contractors. Mr. Onodera and Mr. Ishikawa both suffered severe damage due to the Great Earthquake and hoped to work together to pave the way for the *mirai* (the future), of the shipbuilding industry in Kesennuma. As a result, they played a role of glue in bringing the four shipbuilding companies together. The third company is K-Yard, which moved to Hokkaido with its employees immediately after the Great Earthquake and now runs a shipbuilding company in Nemuro City in Hokkaido[4]. As the company with specialty in building small aluminum boats had many customers in Hokkaido, and was looking to an offer from a customer to use a vacant factory in Nemuro. After the Great Earthquake, the mother factory in Kesennuma became unusable, and the company decided to move its operation to Nemuro. However, the president hoped to return to his hometown of Kesennuma someday, and participated in the establishment of MIRAI SHIPS INC.

So, let's begin the story of the establishment of the new shipyard by the seven companies who came together in quest for a new beginning. This is a story of business owners who recovered from the unprecedented disaster of the Great East Japan Earthquake, and it is also the story of the revival of

3 The representative of Onodera Iron Works is Mr. Shigemasa Onodera. The company was founded in 1949. It maintains, develops, manufactures and installs engines and equipment of boats. The representative of Ishikawadenso was Mr. Yuto Ishikawa in the period between the Great Earthquake and the restoration of the company.

4 The representative of K-Yard is Mr. Masayuki Fujiki. The company was founded in 1974. It builds and maintains small aluminum boats. The headquarters of the company in Kesennuma was destroyed in the wake of the disaster, and was relocated to Nemuro in Hokkaido with employees. He participated in the establishment of MIRAI SHIPS INC. in anticipation of returning to Kesennuma in the future. He also serves as an outside director of MIRAI SHIPS INC.

Heavy oil, which spilled into Kesennuma Bay, and scattered debris ignited and produced a wide flame zone.

local areas and of the shipbuilding industry in Japan.

The Reconstruction Committee for Fishing Port Functions

Kesennuma suffered extensive damage due to the Great Earthquake. The fire caused by the destruction of fuel tanks and the subsequent heavy oil spill into the sea, as well as the image of the ships that were swept inland into the center of the city, are still vivid and remain indelibly inscribed in our memory. Two weeks after the disaster, about 30 companies related to the shipbuilding industry could finally gather to confirm their safety and each other's status. Among these approximately 30 companies were shipbuilding companies, machinery and equipment-related companies and electrical equipment companies, as well as various other companies such as fuel-related companies, food-related companies and fishing gear companies. At the meeting, talks about establishing a committee for restoration and reconstruction were held, and subsequently the "Kesennuma Reconstruction Committee for Fishing Port Functions" was formed.

The purpose of the committee was to rebuild the shipbuilding and fishery industries and the ports, and the immediate goal was to accept bonito vessels which would enter the port for fishery starting mid-June. Once a bonito

Dai-juhachi Kyotokumaru was moved to the city area 750 meters away from the port. It was dismantled in October 2013.

vessel arrives at the port, various tasks need to be done including tasks of landing fish, repairing the boats, loading food for departure, and loading fuel. The shipbuilding industry, which had flourished along with the fish processing industry as mentioned earlier, never wanted to give up their businesses in excuse of the disaster. They had to somehow be ready and in time for the bonito season. This was the unanimous thought of all companies of the shipbuilding industry.

For this reason, work on removal of large fishing boats that had been beached, on securing power supply by installing generators, on repairing flooded electric motors, including repairing damaged engines and equipment of vessels, along with the restoration of lifelines such as electricity and water supply, and preparations, were made to restore the factories one by one.

At the time of this crisis which could be said as a national crisis, various support activities by the government, local governments and public service corporations started in the affected areas immediately after the Great

Earthquake. Among them, the Reconstruction Committee for Fishing Port Functions decided to apply for the "Restoration and Maintenance Project Subsidy for Facilities of Small and Medium Enterprise Groups" (Group Subsidy) of the Small and Medium Enterprise Agency, with the cooperation and guidance of the Maritime Bureau of the Ministry of Land, Infrastructure and Transport.

By filing this application by 58 companies as a group on June 24, 2011, the Committee was able to secure a project cost of approximately 3 billion yen for an estimated total damage of 12.4 billion yen. This subsidy system was designed to assist any group made of small and medium enterprises, which constitute the "core of the regional economy," that could play a leading role in the reconstruction after the Great East Japan Earthquake, by providing financial assistance for restoration of damaged facilities and equipment, once their reconstruction plan is approved by the prefectural government. Their application was accepted in the very first round of the process of this subsidy system.

Restoration support during this period can be mostly characterized by support for temporary restoration. When the damage was minor and could be tentatively restored with some minimum repairs, smooth and quick restoration process was achieved in some cases as the subsidy was admitted even for cases where repair work was already on-going. Conversely, reconstruction of a new building took two to three years or even four to five years depending on the conditions, as the debris had to be removed first.

Nevertheless, the Reconstruction Committee for Fishing Port Functions was on the whole able to carry out immediate restoration activities by getting these subsidies. At the same time, the relationship with the Ministry of Land, Infrastructure, Transport and Tourism (MLIT) actually paved the way to the birth of MIRAI SHIPS INC.

Land Subsidence

Fortunately, there were no human casualties among the employees of the shipbuilding companies. Probably because all of them, with full knowledge about dangers of the sea, were able to take prompt evacuation actions immediately after the earthquake struck. However, facilities and equipment of the shipyards were severely damaged, and restoration had to be started from removal of stranded boats on their premises. This work alone took two to three months, and restoration of facilities and equipment was started thereafter.

It was in August when Kidoura Shipyard was able to accept the first boat. Kesennuma port as a whole, however, was able to prepare in time for the entry of bonito fishing boats. Even in the year of 2011 when the quake occurred, the port was able to defend its position as the number one port in Japan for bonito catches.

Despite the quick recovery, however, the shipbuilding companies became aware that they could not return to the status quo in pre-quake times due to a serious problem—extensive damage by severe land subsidence. The earthquake had caused land subsidence by about one meter, and as a result, about ten meters of the shipyard sites had been eroded by the sea. Therefore, a boat, which until then, had landed beautifully from the bow to the stern when raised for repairs onto the building berth, sank in the sea at the stern. This damage was of catastrophic proportions and not something that could be fixed by construction.

Even if the boat to be repaired could be lifted onto the berth, the work had to be carried out with the stern submerged in the sea, which placed heavy burden on workers. In addition, the land subsidence made work area much smaller, and therefore large boats could not be repaired. It is still the case even in January 2019 at the time of writing manuscripts of this book.

Nevertheless, the shipbuilding industry was extremely busy to cope with the growing demand for reconstruction shortly after the Great Earthquake. Companies had no spare time due to repair and reconstruction work for damaged boats. However, this would not last forever. It was clear that if they continued to work using their short building berths, their work would some-day be inevitably taken over by other shipbuilding companies outside Kesennuma, while they would keep facing an internal diminishing equilibrium. Given that the shipbuilding industry would be essential for Kesennuma, leaving the situation as is was never their option. It was absolutely necessary that a new shipyard capable of handling large ships would be built at a different location somewhere in Kesennuma. This tremendously big project would not be accomplished if each shipbuilding company took on the challenge single-handedly without any collaboration. Unified engagement was definitely needed—that was a common thought that shipbuilding companies started to bear in their mind by the summer of 2011.

Kesennuma Shipbuilding and Marine Industries Reconstruction Council

As mentioned earlier, the Kesennuma Reconstruction Committee for Fishing Port Functions was a group consisting of various companies relating to the fishing port, but the entity was not necessarily suitable to discuss the issue of the shipyards reconstruction. Therefore, once the season of landing bonito from boats was over, members keen to the shipyards reconstruction kicked off the Kesennuma Shipbuilding and Marine Industries Reconstruction Council. It was in July 2011. The council was participated by 32-companies' business owners who are deeply tied up with shipyards via their sweaty hard work.

On August 8, the council submitted a paper of request to Kesennuma City, titled "Restoration, Reconstruction and Emergence of Shipbuilding and Marine Industries in Kesennuma," which led the City to make its official announcement on October 7 regarding the shipyard reconstruction as one of the top priority projects of Kesennuma.

The council members started formulating their own unique vision for a new shipyard. At this juncture, they came up with a plan which can be comparable to a huge shopping center, where shipbuilding companies sit as the core and various specialized companies such as machinery maintenance, electrical equipment, painting, fishing equipment and food companies, gathered together and thus they offer a one-stop service in the shipyard. The plan was to establish a new holding company, and the company would operate the new shipyard that would benefit every member company.

However, a large investment must be done to realize this project, and its finance was a major hurdle. The council never knew how to solve it. The council had a chance to get such information that MLIT would come up with political measures for the reconstruction of the shipbuilding industry, but they found it yet to be ready. One day the council learned about the Facility Maintenance Support Project for Affected Small and Medium Enterprises (a loan facility based on a business upgrade scheme), designed by Small or Medium Enterprise Agency (SME) under the Ministry of Economy, Trade and Industry (METI) and to be executed by Miyagi Prefecture. When they explored it more in detail, they came across a condition that the project money must be guaranteed by some of the main members of applying body.

So the council asked its members if someone raises hands for the joint guarantors, the seven members I introduced in the beginning of this chapter said yes and they would become the founders of MIRAI SHIPS INC. later on. Although the council gave up SME's "business upgrade scheme" in the end because they were not able to fulfill all conditions, this course of events delivered the seven core members of making MIRAI SHIPS INC.

Mr. Kidoura, Mr. Yoshida, Mr. Kosaba, and Mr. Sawada, the presidents of the four shipbuilding companies made up their minds to retain the shipyard by all means, and Mr. Onodera and Mr. Ishikawa were determined to follow it through as they knew that there would be no work at all without a shipyard. In addition, Mr. Fujiki of K-Yard, who had dreamt of returning to Kesennuma someday, decided to participate in the project to establish a new shipyard.

Now in the project, Kesennnuma's shipbuilding business owners became categorized into three types. Firstly, those who intended to create a new shipyard at their own risk to reconstruct the shipbuilding industry in Kesennuma (they are the seven business owners who are the central figures of this book). Secondly, those who supported the construction of a new shipyard and worked towards realizing the construction of the Kesennuma Shipbuilding Complex. Finally, those who could not wait for the establishment of a new shipyard which would happen several years down the line, and had no choice but to walk a different path. Some people chose new paths including relocation and closure of their companies, due to the severe damage caused by the Great Earthquake. The quake continued to ruthlessly shake the fate of the people.

The seven founders thought that making a company by the seven would accelerate their business much faster than walking with other many stakeholders on consensus basis. At that point of time, they envisioned a structure in which each shipbuilding company would be affiliated through the holding company system. With this system Kidoura Shipyard, Yoshida Shipyard, Kosaba Shipyard and Sawada Shipyard could continue as before. "We would not have been able to reach any agreement if we work on a merger from the very beginning," says Mr. Takeyoshi Kidoura.

Each company had a long history. Kidoura Shipyard was founded in 1932, Yoshida Shipyard in 1916, Kosaba Shipyard in 1910, and Sawada Shipyard in 1954. Sawada Shipyard was relatively new, but still had a history of more than 60 years. Certainly, it did not seem likely that these traditional companies would merge that easily. Therefore, they started to study a plan where the four shipbuilding companies are affiliated under a holding company and with collaboration with the Kesennuma Shipbuilding Complex Cooperative.

The Kesennuma Shipbuilding Complex Cooperative

The Kesennuma Shipbuilding Complex Cooperative was established in April 2013. The organization included eighteen companies relating to shipbuilding. In a way, it can be said that the period from 2012 to 2013 was a runway period for the establishment of the new shipyard. The following is a brief history. MLIT established the Project for the Development of Tsunami Rehabilitation Bases in 2011 in response to the enormous damage from the tsunami caused by the Great Earthquake. This project aimed to develop urban areas as bases for reconstruction with the intention of preparing for damage from any future disasters, by using the Reconstruction Grant provided by the Reconstruction Agency. This project aimed to revive the affected areas promptly by taking measures, including for example, relocating major industries to elevated ground. When it comes to the establishment of the new shipyard, such option was discussed that Kesennuma City would use the Reconstruction Grant to secure and reclaim a land for building a construction site.

Meanwhile, the "Reconstruction Support Project Subsidy for the Shipbuilding Industry" of MLIT, was to be used to construct new shipbuilding facilities. This subsidy system was established in 2013 in order to provide full support to shipbuilding companies who had resumed operations by their emergency measures but were still facing difficulties due to the impact of land subsidence among other things. The budget was about 16 billion yen and the subsidy rate was two-thirds (with an upper limit of 8 billion yen). MLIT started accepting applications for this subsidy in August 2013, four months after the formation of the Kesennuma Shipbuilding Complex Cooperative. At this time, the Nippon Foundation[5] was selected from public entries as a supporting organization of miscellaneous management relating to the funds.

As the conditions to be qualified to the subsidy were getting clarified, the governance of the Kesennuma Shipbuilding Complex Cooperative became also more organized.

The seven founders constituted the Board of Directors of the Cooperative. At first, they planned to make the Cooperative as the organization that would receive the Reconstruction Support Project Subsidy for the Shipbuilding Industry. Even at this stage, the vision of the new shipyard remained something like a shopping mall complex. However, the idea had to be dismissed because they found that the Cooperative, which is not a legal entity, was not eligible in applying for the subsidies.

Eventually, they chose to establish a new holding company, MIRAI SHIPS INC., which was to become the owner of the new shipyard. It was assumed that the seven business owners as founders would take responsibility of running the new company, and their existing companies were to be affiliated to the holding company respectively. Every member company under the Cooperative would able to commonly join and utilize the new shipyard by paying the rent fee.

However, it took a great amount of time to convert the plan into reality. What should the new shipyard look like? What about the structure of the new company, getting subsidies, finance, or pay-back plan, and so on? The

---

5   The Nippon Foundation is a public interest incorporated foundation that supports social welfare, education, culture, ships and oceans, and overseas cooperation assistance projects, using a portion of the proceeds from boat racing as a source of revenue.

more they dug into the plan, the more they saw problems to be solved piling up.

Deloitte Tohmatsu

Kesennuma City and the Kesennuma Chamber of Commerce and Industry sponsored a training school called the Future Management School, an institution for the development of human resources, aimed at fostering business managers. Mr. Takeyoshi Kidoura and Mr. Shigemasa Onodera enrolled as the first-generation students of this school, which started in April 2013. They intended to incorporate their new vision of the shipbuilding industry in Kesennuma into their business plan. At this time, Mr. Kidoura tried to find ways to diversify the shipbuilding industry, and Mr. Onodera decided to work on a plan to construct a new shipyard.

This school also co-sponsored by the Tohoku Future Creation Initiative, a collaborative initiative by employer associations, universities, governments, and civil society organizations to support reconstruction in Tohoku[6] and Deloitte Tohmatsu[7], also supported this initiative and extended their cooperation. Deloitte Tohmatsu partners and staff were involved in the Future Management School as mentors (instructors/advisors) in order to enhance the business planning capabilities of students. Mr. Gentoku Yoshikawa and Mr. Eiji Sato of Deloitte Tohmatsu became acquainted with Mr. Kidoura and Mr. Onodera as their mentors.

---

6 The Tohoku Future Creation Initiative was an organization that ran the Human Resource Development School for the purpose of nurturing human resources. It was established through collaboration between industry and academia aiming for independent and creative reconstruction in the areas affected by the Great East Japan Earthquake. The representative founders were Seiichi Ohtaki, a professor at the Graduate School of Economics and Management of Tohoku University, and Kentaro Ohyama, the Representative Executive President of Iris Ohyama (a houseware company) (both titles are those at that time). Its various five-year activities were completed on March 30, 2017.

7 The Deloitte Tohmatsu Group belongs to Deloitte, the world's largest network of comprehensive professional services firms. Group corporations include Deloitte Touche Tohmatsu LLC, Deloitte Tohmatsu Consulting LLC, Deloitte Tohmatsu Financial Advisory LLC and Deloitte Tohmatsu Tax Co.

As the school was a six-month limited period institution, there was not enough time to complete a perfect business plan. Nevertheless, the business plan created by Mr. Onodera later became the conceptual model of MIRAI SHIPS INC. On the other hand, Mr. Kidoura fell ill in September 2013. He was diagnosed with cerebral infarction and was forced to stay in the hospital for a long time.

When Mr. Yoshikawa visited the hospital where Mr. Kidoura was in, he found a document on the bed—a summary of tasks to realize the new ship-yard—prepared by MLIT. Mr. Kidoura told Mr. Yoshikawa to read through it and asked him for help with this task. Mr. Yoshikawa was amazed by the offer, wondering what Mr. Kidoura was doing despite his critical situation of suffering from a life-threatening illness, and he immediately accepted the offer.

Deloitte Tohmatsu had its local offices in the three affected prefectures of Iwate, Miyagi and Fukushima, and they started their support activities for reconstruction immediately after the Great Earthquake. In fact, Mr. Yoshikawa had visited Kesennuma more than half a year prior to the start of the Future Management School, and already had some knowledge of recon-struction of the fish processing industry. After that, Mr. Yoshikawa and Mr. Sato played the role of a mentor, guiding the seven business owners to the launch of MIRAI SHIPS INC. by giving them advice at key points.

Mr. Onodera's business plan reflects his own views and thoughts, but it was never shared by any founding members except Mr. Kidoura, in the initial stages. It is no wonder that everyone differed in opinions, and it took very long time to integrate all views into one plan. Every member was proudly running his own company, and things did not move forward once anyone said no.

Mr. Yoshikawa recalls, "All founding members faced such challenges that none of them ever imagined nor experienced before by this establishing a state-of-the-art shipyard with a huge amount of money. In addition, it was necessary for this task to be agreed on by all members, members who, till then, got used to make his own decision about all kinds of managerial mat-ters on their own risk and judgment. It was bound to take time because they were unfamiliar with collaborative work." With the talented mentors at the helm, the vision of the future of MIRAI SHIPS INC. gradually began to take shape.

One More Supporter

Along with Mr. Yoshikawa, there was one more person who continued to advise the seven business owners helping them in all ways. This individual was Mr. Akira Sueishi who played a central role in this project at Kesennuma City Hall. In October 2013, when the first students of the Future Management School graduated, Mr. Sueishi was seconded from MLIT to Kesennuma City Hall. Hs role was a rescuer to Kesennuma City Hall, which was getting dysfunctional due to increasing reconstruction-related duties in addition to the normal duties. Eventually, Mr. Sueishi was given his special mission to support the MIRAI SHIPS INC. project.

Mr. Sueishi's first job right after his secondment was to convince the Reconstruction Agency that the shipyard project is applicable to the Reconstruction Grant. This was quite a difficult task, because, in their view, shipyards, which are always located along the sea, could be easily destroyed if a tsunami occurs again no matter how robustly reconstructed. Given this, the Reconstruction Agency had a hard time to regard a shipyard as an effective reconstruction base.

To counter those views, Mr. Sueishi looked to a shipbuilding facility adopting the shiplift system as a countermeasure to tsunami damage. The shiplift is a system to lift up ships from the sea like an elevator and to move them to a flat workplace to start repair work. With a single shiplift, it is possible to lift up several ships and to move them to the workspace inside the tide embankment. It is safer and more efficient compared to the conventional slipway system because the sea and the workspace are separated.

Needless to say, ships are indispensable to Kesennuma, a fishery town. It is very obvious that if an earthquake disaster strikes again, the fishing industry will not revive if there are no ships at all. For ships at sea, danger is unavoidable in the event of a tsunami. However, even in the event ships are damaged, the town can be revived as long as it has a place and a company to repair those damaged ships. Therefore, a shipbuilding facility equipped with measures for countering tsunami damage is absolutely necessary.

Ship retracted

Ship lifted

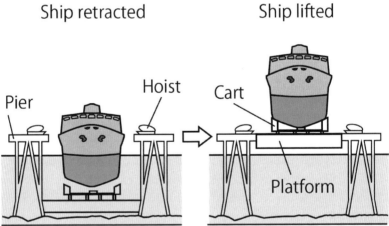

Pier

Hoist

Cart

Platform

Shiplift system. The ship is lifted to the sea level by the shiplift that protrudes into the sea and is stored in the shipyard. The shipyard is separated by the high tide embankment. The tide embankment opens and closes at the site of the shiplift and ensures safety.

With this logic, Mr. Sueishi was able to successfully convince the Reconstruction Agency that the planned shipbuilding facility would be dealt as a legitimate reconstruction base, and it was subsequently decided that Kesennuma City would secure and ready a construction site to build a new shipyard. Thereafter, Mr. Sueishi devoted himself to supporting Mr. Kidoura and the others for the successful acquisition of the Reconstruction Support Project Subsidy for the Shipbuilding Industry for MIRAI SHIPS INC., and to finishing the shipyard without any hindrances.

A Holding Company or a Merger?

A large-scale reconstruction project, costing more than 10 billion yen for construction of the shipyard alone and approximately 20 billion yen including development of the construction site, was gradually getting shape.

In fixing the specifications of the new shipyard, heated discussions were held to ascertain which equipment was required and which not. Even after the specifications were set, they reconsidered whether the equipment was essential or not in light of burgeoning construction costs. It was found that the cost of installing the shiplift was unexpectedly high due to problems of the seabed and the land bases. It took time to solve these issues one by one. Mr. Yoshikawa and Mr. Sato from Deloitte Tohmatsu and Mr. Sueishi from MLIT participated in the meetings, which were usually held once every two weeks, sometimes on-site and sometimes through videoconferencing, and led the discussions on these issues.

Another important issue was the structure of the new company. Mr. Yoshikawa and his team were involved in the process of making the business plan for the new company, and advised their views that the option of holding company could not produce expected integration benefit. Making a new holding company while retaining the four existing companies could rather add cost. The holding company method can be effective when, for example, affiliated companies have their own brands and do business with a respective unique customer base. But assuming that the four existing companies are to be engaged in similar work in the same business area, it would be difficult to make the holding company system effective.

Mr. Yoshikawa and his team conducted long and thorough deliberations on this issue. Mr. Yoshida was not persuaded at all at the beginning because he had lots of uncertainty towards the future of his company including aging employees and unresolved successor issue, but he ended up with accepting Mr. Yoshikawa's persistent explanations. Mr. Yoshida then persuaded Mr. Sawada, another opponent of merger option, which paved the way to all agreeing with the merger of the four companies. In April 2014, the merger agreement was signed.

On this issue, Mr. Kidoura recalls, "I think we had a common understanding that the four companies are destined to be merged someday, say, a hundred years down the line. But when asked about time after five or ten years, the each view does differ. For example, someone wants to run his company for another five years and may think that the merger will be OK after that, or, another wants to run for about 15 years and may hope that it will remain as is. It depends. However, thanks to the sincere advice from Mr. Yoshikawa, all of us could get fully convinced that, if we wish future succeeding managers after us will be able to keep and run business, in stable manner we have to merge and make one company now."

Mr. Kidoura was born in 1970 and is by far the youngest among the seven founders. He would be only 49 years old when the new shipyard was built. He is a decade or more younger than Mr. Tsutomu Sawada or Mr. Keigo Yoshida. Mr. Kidoura wanted to adopt an aggressive approach to his business and would have rebuilt the shipyard armed with state-of-the-art equipment unless the Great Earthquake occurred. On the other hand, Mr. Sawada and Mr. Yoshida are the kind of people who are familiar with every detail of the shipbuilding industry given their long history and have also been party to heated arguments with their peers in the past. It was difficult for people of this generation to part with their company which they have nurtured for long, and to agree to a merger easily. It was an agonizing choice interlaced with conflicting thoughts and mixed emotions.

Environmental Concerns

With the backing of Deloitte Tohmatsu, MLIT, and Kesennuma City among other supporters, the shape of MIRAI SHIPS INC. was gradually starting to emerge. However, a new issue erupted unexpectedly. The question was posed, saying that construction of the new shipyard was unacceptable from the perspective of environmental issues.

Asahi-cho, the site where MIRAI SHIPS INC. was to be built, had been selected from several candidate sites. Asahi-cho was selected because it had a water depth of more than 10 meters and because it required no landfill, leading to a short construction period and reduced construction costs. The choice had an aspect of effective use of the damaged land, where it had fuel storage facilities that were destroyed by the Great Earthquake.

However, when this construction plan was covered in local newspapers in February 2013, the fish processing industry submitted a request to the Mayor of Kesennuma City.

The request stated that the plan to build a shipyard was proceeding without any explanations to fish processing companies. Further, if a shipyard were to be built in Asahi-cho, they have concerns about how much it would impact aspects of hygiene relevant to nearby fish processing companies and concerns about a consequent decline of the fish processing industry. Requests were also submitted to the Kesennuma City Council, the Prefectural Governor, MLIT, the Fisheries Agency and the Reconstruction Agency.

However, as shipyards and fish processing factories have been existing next to each other for a long time since before the Great Earthquake, it was hard to think that a new adjacency in Asahi-cho would lead to environmental degradation. Nevertheless, 80 percent of manufacturers in Kesennuma City engage in the fish processing industry, and they have great influence. Moreover, they also posted opinion advertisements in local newspapers twice.

In response to this, MIRAI SHIPS INC. and Kesennuma City took actions such as holding briefing sessions. Although there were no legal restrictions that could hinder the shipyard construction, what was most important was ensuring coexistence with the fish processing companies. The top priority was on the reconstruction of Kesennuma, and any confrontations were out of the question at all.

In October 2014, Kesennuma City established the Environmental Measures Evaluation Committee[8] (with Professor Osamu Nishimura of

---

8   The objectives of the Environmental Measures Evaluation Committee were to

the Graduate School of Tohoku University as Chairman) and formulated a system to evaluate the environmental measures of MIRAI SHIPS INC. MIRAI SHIPS INC. also drafted a construction plan once again taking into consideration environmental measures. The plan included as many environmental measures as were possible, for example, reduction and attenuation of work noise, installing water collection equipment to prevent hull wash wastewater from flowing directly into the sea, putting up dust-proof nets to prevent paint and other things from scattering, and introducing an environmental management system. These measures resulted in making the shipyard more environment friendly, which we can call a happy by-product.

As a result of these measures, the turmoil got gradually subsided. We will see its consequence when MIRAI SHIPS INC. starts operations.

The Birth of MIRAI SHIPS INC.

MIRAI SHIPS INC. came into existence on May 1, 2015. On May 26, the Reconstruction Support Project Subsidy for the Shipbuilding Industry became granted. Significant progress has been made toward the construction of the new shipyard. When the provision of the subsidy was decided, all the seven members of MIRAI SHIPS INC. visited the Nippon Foundation and humbly received the notice of determination from the president. It was quite an event and was covered and broadcast by a local TV media.

Even at this stage, only the basic frame of the company had been created. The four companies had not merged yet, and it was necessary to calculate the merger ratio in order to merge. Instead of directly determining the definitive composition ratio by actual calculation, a provisional calculation was made so that the members could follow and understand logic and impact of the numbers. In this process, each business owner came to know raw financial data of other companies that was never shared before, which consequently ensured complete transparency among them.

---

validate the environmental measures, which were going to be implemented by the shipbuilding company, for the shipbuilding facility to be constructed in Asahi-cho District, and to conduct monitoring surveys after the start of the facility, in order to ensure that the shipbuilding company is implementing the environmental measures and to promote understanding of citizens regarding the shipbuilding facility.

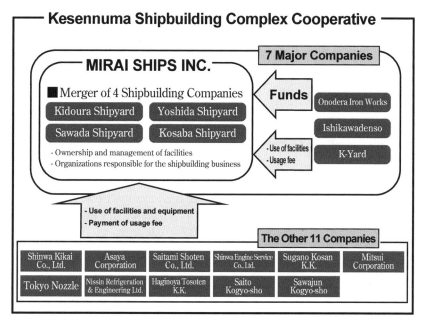

The relationship between MIRAI SHIPS INC. and the Kesennuma Shipbuilding Complex Cooperative

The provisional calculation was conducted by Deloitte Tohmatsu and shared to all members accordingly.

Based on this report, Mr. Sawada made one proposal, urging all four companies to clean up all negative assets to be well prepared for the merger. He insisted that success of the new company will never be guaranteed unless it is financially healthy when started.

His view was no doubt coming from the profound impact of the Great Earthquake. Although the Great Earthquake generated significant reconstruction-related demand, the shipowners ordering those work also suffered profound damages. Moreover, the nature of the relationship between shipbuilding companies and shipowners has traditionally been something where the shipowners at client position were overwhelmingly stronger than the shipbuilding companies.

Mr. Sueishi notes that, "Although both of shipowners and shipbuilding companies suffered heavy damage from the Great Earthquake, shipbuilding companies had to bear more burdens due to the power dynamics. Mr. Kidoura, in particular, had to accept very low prices, which fairly impacted his company financially." More or less same thing went with other companies besides Kidoura Shipyard.

This problem arose from the business structure of the shipbuilding industry in Kesennuma. From shipbuilding companies to other companies, such as machinery, equipment, electrical equipment and painting companies, every company get ordered directly from shipowners and deliver their own specific work, and it is difficult for them to see from total process management viewpoint. For example, painting comes after welding, but how to proceed the two process heavily depends on tacit harmonization between two companies. Although problems seldom occur because those involved are highly professional, it is also a fact that lack of process management led to bearing unnecessary costs, and once the amount of work gets overflowed, it turns into weakness.

Taking this opportunity, the members of the four companies began to review past business practices and to conduct thorough process control consciously. They also made efforts to improve process control by learning from leading Japanese automobile manufacturers. Mr. Sueishi says, "I thought a lot of changes occurred in their mindsets. In the past, they undertook work from shipowners as they say solely because of their longstanding relationship. I heard that sometimes, they could not charge an extra fee even when some specification was changed after the completion of work. I remember Mr. Kidoura mentioning that he learned that specifications must be elaborately discussed with the shipowner before accepting an order, and then he became able to run his company more stably."

In the end, Mr. Sawada's proposal resulted in improved unification and enhanced management ability of the four companies. The companies began to handle tasks jointly, and work together to eliminate the debt of each company. The effect of the preliminary calculation was enormous. There was more mutual transparency in place, and they were now fully ready to tackle the business head on. They all believed and committed that there would be no fall back and that there would be no way but to walk on the path towards the establishment of the new shipbuilding company—a journey which they had already embarked on together.

In addition, Onodera Iron Works, Ishikawadenso, and K-Yard were now in a position to pay charges for using the facilities of the new shipyard while investing in MIRAI SHIPS INC. Similarly, members of the Kesennuma Shipbuilding Complex Cooperative would also pay charges to the MIRAI SHIPS INC. for using its equipment and facilities. As for the directors of the Cooperative, as the four shipbuilding companies have merged, all have already retired, except for one president.

Mirai in Your Hands

Let's turn back time a little. After the "loan facility based on the business upgrade scheme" managed by Miyagi Prefecture did not go well, the seven members of MIRAI SHIPS INC. were very worried about finance. Finance by government subsidies was essential, but they knew that they would have to show their own money anyway. An enormous amount of money.

From May 2014, they visited several banks to explain their business plan and ask for loans. Recalling the situation at that time, Mr. Onodera says, "My impression was that banks did not positively evaluate our plan, and getting loans might be tough anyway." In hindsight, the truth was that in fact banks took the plan as immature and detailed examination for loans was yet to be done. The seven members further refined the business plan and came back to banks with loan request in October 201. The situation then started move forward. In February of the subsequent year 2015, a loan syndicate consisting of eight banks led by Mizuho Bank[9] was established. In March, MIRAI SHIPS INC. managed to submit an application for the subsidy to the Nippon Foundation in the nick of time. This was the course of events for how the subsidy came to be granted.

In April 2018, MIRAI SHIPS INC. merged with the four companies, Yoshida Shipyard, Kidoura Shipyard, Kosaba Shipyard and Sawada Shipyard, and physically became one company. The four ex-owners became senior managers on the same one boat and embarked their new voyage.

---

9   A loan syndicate is a group of financial institutions which is formed to provide a syndicated loan.

The business plan listed "efficiency in operations," "establishment of a rapid and efficient counting management system" and "transition from the separate ordering system to the collective ordering system" as their immediate managerial issues. The intension of the list is to reform their old business practices as described above, to review operations relying on their experience and guts-feeling, and to shift to a more efficient management system. Another issue listed in the plan was the "aging in human resources." Employees with high skill in shipbuilding and repairing techniques and accumulated know-how was rapidly aging. Their techniques and know-how must be succeeded down to the younger generation. Those key issues including younger workforces will come down to the ultimate objective of MIRAI SHIPS INC., which is to ensure the steady and continued growth of the shipbuilding industry in Kesennuma and to revitalize the town of Kesennuma.

Shortly before the Start of Operations of the New Shipyard

The new factory will be completed in April 2019. President Kidoura, spending very busy life every day, has one big concern. It is the untouched issue of the integration and unification on employees' side. Managers are now unified after overcoming lots of challenges altogether, but employees are yet to be unified.

Even at this stage, after a series of interviews have been completed, the names "Kidoura Factory," "Yoshida Factory," "Sawada Factory" and "Kosaba Factory" still remain, and each workplace seems to be operated independently. Certainly, work is gradually being shared among the factories; for example, ships designed at the Yoshida Factory are being built at the Kidoura Factory.

However, the customers the employees meet directly are basically the same as in the past, and the working environment has also not changed. Greeting to colleagues from other company in a shiny, new locker room or the relaxed space of restrooms seems to still be a long way off. President Kidoura is concerned that, because of his busy days, he has not been able to create occasions to bring them together.

However, he looks optimistic in a sense towards the situation. He says, "We are in the same business, but came from different companies, which may or may not make relationships a bit awkward at first, but I hope that Mr. Onodera and Mr. Ishikawa, who have long-standing relationships with the employees of all shipyards, will play a supporting role."

The shipbuilding workspace belongs to craftsmen. The craftsmen can only work in the way want to and they did in the past. Mr. Kidoura believes that enforcement will not work. "Perhaps there is no all-encompassing solution. I think it is important to build on every achievement one by one."

Keeping a natural posture even in management—that is Mr. Kidoura's style.

"Of course, I would like to hold a grand opening ceremony to show employees that the project has been supported by so many people. I think the best gift for them is to let them know that people around have high hopes from us."

*

A new highway, the Sanriku Coastal Road, is scheduled to be constructed just above the new shipyard. Piers for it are already going to be completed. Although the plan had been in existence for a long time the Sanriku Expressway could not have been fully constructed so quickly without the Great Earthquake. There seems to be a sense of anticipation that a big disaster has brought about an "impossible future," a future which would have never happened otherwise. MIRAI SHIPS INC. may be one of such "impossible future." A future that seemed impossible can be created by integrating power from many supporters. The members of MIRAI SHIPS INC. may stuck with problems or get lost, but they will surely drive their motivation for tomorrow while keeping candid exchange of various opinions and aspirations.

To Pass on the Historic Shipbuilding Industry of Kesennuma to Future Generations of *Mirai*

Akira Sueishi

International Affairs Office, Shipbuilding and Ship Machinery Division, Maritime Bureau, the Ministry of Land, Infrastructure, Transport and Tourism

Former Vice-Counselor of the Industrial Revitalization Strategy Division, Department of Industry, Kesennuma City Hall

I was seconded to Kesennuma City Hall from MLIT from October 2013 to the end of June 2015. When I arrived, more than two years had passed since the Great Earthquake, but the scenes of the town were far from recovery. I tried to rent an apartment to live in but in vain. I got one of temporary houses for victims 10 kilometers away from the center of the town and commuted by car.

My mission was to work for resurrection of shipbuilding industry in Kesennuma. More concretely, the mission was to build a new shipyard in Asahi-cho and to make the damaged four shipbuilding companies revive as the new company, that is, MIRAI SHIPS INC. As a city official in charge, I did my best to ensure its finance by getting Reconstruction Grant from the Reconstruction Agency and to secure the site for construction as my critical mission, but the most critical issue, to my memory, was whether the concept of MIRAI SHIPS INC. would happen or not.

I came to be acquainted with Mr. Kidoura and Mr. Onodera first, both enrolled in the Future Management School to elaborate their plans about making new shipbuilding company. From time to time, they came to me for my advice and our discussions covered various issues with other five key person involved. When it comes to the crucial phase, the seven people always sit down and discussed face to face basis. They often failed to reach an agreement because of their individual views and opinions. In such an occasion, I went to Mr. Yoshikawa, a good friend of mine by then, to get his advice.

I saw such moments several times that the talk would be broken off. The most critical one was when the tentative numerical simulation was shown. The poor financial performance of Mr. Kidoura's company made me so worried because the fact could undermine his pro-merger position. After long discussion, Mr. Sawada, who had kept his negative position against merger, eventually came to agree and that was a major turning point.

The new shipyard will be completed soon. It will be a state-of-the-art large scale of shipyard and they will be able to build bigger fish boats. I hear voices wanting to see revival of large scale tuna boats which used to be built in Kesennnuma. I want to see them to build Coast Guard ships. They will also be able to build large size equipment for offshore wind power generation using their advanced shiplift facility. They will be able to take orders from oversea markets. So many opportunities await them.

The new company and facility have been made by such a large size investment that even a big shipbuilding company might have been prudent to make. I would like them to work hard from now on for the sake of future of Kesennnuma.

Bringing Happiness to More People as a Symbol of the Revival of Kesennuma

Mitsuyuki Unno

Managing Director of the Nippon Foundation

The Nippon Foundation supports a variety of public-interest projects in various fields including social welfare, education and culture, marine vessels and oceans, and overseas cooperation aid, by using a portion of the proceeds from boat racing as our source of revenue. The Foundation used to be named as the Japan Shipbuilding Industry Foundation, and since its foundation, our main focus has been naturally on supporting shipbuilding-related industries, which used to be very essential industry in Japan. Because of this background, we have special sense of mission that we must take immediate actions in the event of marine related disaster in particular, amongst our numerous social contribution activities. And the tsunami damage caused by the Great East Japan Earthquake posed the greatest challenge to us.

In that context, MLIT chose us an entity to manage their Reconstruction Support Project Subsidy. And we started our own support activities for reconstruction in the early stages post the quake. We entered the affected areas immediately after the quake and conducted surveys and provided emergency assistance. We also took the initiative in providing assistance and support for the damaged shipbuilding companies, which were equally helpful to shipbuilding-related companies.

To my memory, it was in July 2013 that we firstly met the members of MIRAI SHIPS INC. We visited Kesennuma as part of a preparatory hearing to ship operating companies in the Tohoku area looking to the public offering for the subsidy staring in August. I remember that the name of the entity they belonged to listed on their business cards was not of MIRAI SHIPS INC. but Kesennuma Shipbuilding Complex Cooperative, which meant to us that the Cooperative would receive the subsidy and build a new shipyard, and member companies would move to the new place altogether. But the plan evolved to making MIRAI SHIPS INC.

We had several occasions to meet key members after then, but to be honest, we often became worried to see their disunited situation in terms of direction. We asked them to change it because we believed that the unity was the minimum requirement for the success of this big project.

The atmosphere changed a lot after the completion of merger in April 2018. I do feel their sense of solidarity as comrades and their willingness to challenge their new business altogether, which makes us very happy.

MIRAI SHIPS INC. will make them a symbol of revival by building new ships one by one. As the shipbuilding industry is an industry with a wide range of bases, the growth of MIRAI SHIPS INC. is directly connected to happiness of the local people and will contribute to creating the future of Kesennuma.

Now that they have reached at the start line now. By fully making use of their new facilities, they will be able to develop various kinds of business. I would like them to work hard with the spirit to be the pioneering role model of revitalization of the shipbuilding industry in Japan, and to go beyond the domestic market into the global market.

2 | THE SEVEN BUSINESS OWNERS

1 | Keigo Yoshida
Representative Director and Chairperson of MIRAI SHIPS INC.

Mr. Keigo Yoshida, the president of Yoshida Shipyard, has been leading the shipbuilding industry in Kesennuma along with Kidoura Shipyard for a long time. In their long history, there was a time when they regarded each other as rivals and competed with each other. This relationship between the two companies changed significantly due to the Great East Japan Earthquake. The shipbuilding industry hit by the catastrophic damage to had to be reconstructed by all means. Mr. Yoshida decided to embark on a new path, to start from scratch, just as Japan had risen from the loss of World War II. He played a major role as an anchor in the challenging project of building a new shipyard.

The Number One Shipyard in Japan in Terms of the Number of Ship Types

(Interviewer: hereafter, I): I heard that Yoshida Shipyard has a history of more than one hundred years since its foundation.

(Keigo Yoshida: Y): Yes, my grandfather Keitaro Yoshida started the shipbuilding business in 1916. At that time, fishing boats were still made of wood. Then so-called "semi-steel boats" which had wooden outer shells in the hull as well as a steel bridge, began to be constructed. Today, all boats we build are "steel boats," with the entire body of the boat being made of steel, as you know. Our business has been from the beginning to build new boats and repairing them. Shipbuilding companies in Kesennuma were merged into Kesennuma Shipbuilding Company, a joint enterprise, which happened in 1942, a war time, and our company joined in it. But in 1946, after the war the merger was broken and fell apart again, and that was the time when Yoshida Shipbuilding Company was established, and it was reorganized and became the current Yoshida Shipyard in 1962.

Our first ship after the reorganization was a cargo ship named Dai-ichi Reibunmaru, built at the request by a shipowner in Hakodate, Hokkaido.

It had a gross tonnage of 499 tons. I was still a boy at that time, but I remember that it was a huge ship. The large size of cargo ship by our company was the first but the last and after this, our domain went to fishing boats.

Keigo Yoshida

I: What was the characteristic feature of Yoshida Shipyard?

Y: We were able to build many types of fishing boats, such as salmon and trout boats, saury boats, tuna long-liners, offshore trawlers and squid boats among others. We also built ferries for a local shipping company Oshima Kisen as well as fireboats and ambulance boats. Our company was the number one shipyard in Japan in terms of the variation of ship types we built. It is experience that counts in shipbuilding. Every ship is different. As we have an experienced design engineer of the wartime generation, we were able to build various types of ships primarily thanks to his basic design framework. Of course, he will remain the key at MIRAI SHIPS INC.

I: I heard that when the four shipbuilding companies merged into MIRAI SHIPS INC., you refused an offer to be the president.

Y: Yoshida Shipyard fell into a financial crisis when my father resided as the president. He was good natured, and cared for shipowners, and tried to be with them as much as possible. In the end, our company could not collect the sales proceeds and suffered severe losses. I took over his presidency in 2002. How could someone with such an experience become the head of the new company? I was very consistent in that sense from the very beginning.

The Burning Sea

I: But I heard that you have brought your business back on the right track and that the company got back on its feet successfully.

Y: Sure, I rebuilt the company over a period of three to four years, and I was relieved that our business had started doing well. However, the Great Earthquake occurred just at that time. When the disaster struck, I was in

Hachinohe in adjacent prefecture for work. There was no transportation to return and even no taxis running. I was stranded for about three days at a gymnasium which was used as an evacuation center. My cell phone ran out of battery and there was no way to get information about the traffic situation, so I had to go back and forth between the station and the gymnasium several times to check whether the trains had started operating or not.

On the third day, a television was brought to the evacuation center, and I saw the video of Kesennuma Bay on fire on TV, which really amazed me. "Why and where on fire?" I learned later that tanks for marine fuel along the coast had collapsed due to the tsunami and the heavy oil spillage had gone up in flames. When I went to the station early in the morning on the fourth day, there were staff members of a bus company and they were saying, "We are operating an extra bus to Tokyo for now." I got on the bus, and disembarked at Ichinoseki Station where I had parked my car. I got in the car and headed to Kesennuma. On reaching there I found that the area from Kesennuma Station to the seaward side had been destroyed by the tsunami and the scenery had changed completely. I had no idea where I was walking. I got to the company somehow, but nobody was there. I didn't even know where my wife was or what had happened to her. I called up my four daughters in Tokyo and conveyed to them, "Be mentally prepared for the worst case."

About a week later, one of my daughters informed me that she had received an email from her mother that she was safe. We all cried in relief at that time. My wife said that she had been hit by the tsunami on the second floor of our house and had been alone there for about two days, and that she went to her sister's place and stayed there. Fortunately, the employees who were working in the company on the day of the quake were also safe, although it took about a month to confirm this finally.

Securing Oil in Kesennuma by Repairing Tankers

I: When did you start the restoration of the shipyard?

Y: The shipyard had completely vanished due to the tsunami, and only the propeller and the propeller shafts, which were to be loaded on a boat that was being built, had not been swept away. The hull, the most important part, was found on the premises of the Kidoura Shipyard next door. The boat's shipowner, who visited us out of concern, said, "Please continue building the boat." Another shipowner, whose ship was being remodeled to extend the length of the hull, said "Please continue the work." So, I had no choice but to gear up and get going.

As I had absolutely no idea at first where my employees were, I posted a notice in front of the office that said, "Get together, as we are starting work." Eventually, the employees came back, and started removing the mud at the shipyard. Next, we restored the equipment. I prepared hot meals for the employees in the daytime, and let them return home before dark.

As I said before, we had a problem of oil storage in Kesennuma because the fuel tanks had been swept away by the tsunami. At that point in time, I received a request from a fuel company, Kesennuma Shokai. Our company had built a *yuso-sen* or an oil tanker[10] named "Dai-ju Kokomaru," for them in 1983, but it had been washed away inside the bay and had burned black. However, when I got on the ship, the inside was intact and the engine could be started. So the president of the company asked us, "Please check our ship. I docked it on the quay of the commercial section of Kesennuma Port. The city of Kesennuma is in trouble as there is nothing to store oil, so please fix this ship quickly for Kesennuma." In order to fix it, the berth and the equipment for mounting the ship on the berth had to be restored on a priority basis. So I asked a salvage company to pull up the rails of berth No. 5, which had been turned upside down by the tsunami, and had them install it in the original location for use. I also explained the situation to Ishikawadenso, provided details, and requested them to rewind the motor of the hoisting winch first, and then the motor was activated. The last part to be repaired was the grounding winch to lift up ships from the sea onto the berth. However, it was difficult to repair it indeed since it was hydraulic. Eventually, the oil tanker was lifted to the berth by mobilizing labor, and repair work was started in earnest.

---

10 A ship equipped with a liquid tank to transport oil and heavy oil. A tanker.

That was May 8. We worked like mad. The jib crane[11] could not travel because the motor of the base was damaged, but the motor of the derricking device was alive. In other words, it could not be moved back and forth, but could be turned and moved up and down. As for electricity, a shipowner who had ordered a new ship contacted a contractor by phone and procured a 100 kilovolt motor generator. With this, the cleanup progressed considerably. The oil tanker was repaired quickly, and this helped in the storage of oil in Kesennuma.

I: Did demand for ship repairing and construction of new ships get a boost after the quake?

Y: As only a few repaired boats could be used, the work of building new ships definitely increased. However, we could not set schedules of delivery dates. For example, a shipowner of a saury boat, which was under construction, wanted it to be ready by August 20 when the saury fishing season begins, because saury can be sold at high prices early in the season. I knew it very well, but I could not deliver the boat on time, and I remember feeling miserable about it.

MIRAI SHIPS INC., A Symbol of Industry Survival

I: Yoshida Shipyard was able to recover quickly. Why, then, are you moving towards the establishment of a new company?

Y: I have only four daughters, no successor. I used to think that it would be okay to get along somewhat in Namiita District[12] under the name of Yoshida Shipyard as long as it was only for my generation. However, I changed my mind two or three years before the quake. I told the employees that I would hand over the business not as an inheritance but to an excellent, meritorious successor, and began to focus on employing young people keeping in mind continuation of my business. That was when the Great Earthquake struck. Meanwhile, many shipbuilding-related

---

11 A crane that carries load by suspending an object (a ship in this case) from the tip of a turning arm (jib).

12 Shipyards in Kesennuma used to be concentrated in Namiita District, along the northeast coast of Kesennuma Bay.

companies, such as painting and machinery companies, were in a difficult situation, and I feared that "the shipbuilding industry in Kesennuma would fail if things continued this way."

The impact was not limited to the shipbuilding industry alone. The fishing industry in Japan itself is losing momentum and is facing shortage of crew for fishing boats. In the present circumstances, no one has spare resources to nurture young people. Many people stopped working in the wake of the quake. So there was a sense of impending crisis that the fishing industry in Kesennuma would decline without the hope of MIRAI SHIPS INC.

Immediately after the quake, the Reconstruction Committee for Fishing Port Functions was established by the initiative of Mr. Onodera of Onodera Iron Works. Business managers from various industries, including shipbuilders, gathered and discussed, and we decided to "do our utmost to temporarily restore the fishing port in cooperation with all industries in order to make the shift before the bonito fishing season." During talks about the current status, each one of us, full-grown adults, were often at a loss for words, unable to continue talking, and trying, unsuccessfully, to hold back the tears. It was in this kind of a situation, that a vision for the establishment of this new company emerged. I too felt, "Let's do our best all together, let's get together and get through this phase for the present." I thought that MIRAI SHIPS INC. was the symbol for that, as it is easier to get together under a banner. At first it was just something as simple as that.

I: What led to the new company take concrete shape?

Y: We came to know that the national government would provide a subsidy for shipbuilding reconstruction, and the vision of establishing a new company suddenly became a realistic option. But at the same time, it made me puzzled. Many elderly people were working in my company, just as in the case of other shipyards. It might be happier for them to work in my company without having to face any changes, than to work in a radically different environment. There were times when I thought of it this way, to be honest.

But I made my mind that this would be a lifetime chance and I would have no option but to take it. It was planned initially that the four ship-building companies would establish a holding company, which would affiliate the four shipbuilding companies. We, the presidents of the four shipyards, also visited the Nippon Foundation, a fund-establishment corporation, and talked with members of the Foundation. Even when they asked us how we would steer our business in the future, I replied, "All four of us here will take command with the spirit that each one of us is the president of MIRAI SHIPS INC." I thought that the new company would not become reality without this spirit.

After that, thanks to the support from Deloitte Tohmatsu, the path toward making a new company was rapidly created and paved. Although it was not yet clear how much subsidy the government would provide and what would be the costs we would have to bear at this point of time, all four of us got united and make this deal completed by all means in order to succeed the shipbuilding industry to the future generation in Kesen-numa. Finally, all the members of the seven companies, consisting of five shipbuilding companies including aluminum boats builder K-Yard and two shipbuilding-related companies, signed the agreement. This was the starting point of MIRAI SHIPS INC.

I: Did you encounter any crisis in the process of making the new company?

Y: There were several occasions when I thought of giving up. The first one was when we worked hard on finance. After enormous time and efforts, it almost came to a point that we could get money from local banks, but it ended up with a failure, which really shocked me.

The second one was when Kesennuma City's plan regarding the relo-cation of the shipbuilding base from the current Namiita District to the port area by making the new company was opposed by the fish process-ing industry by the reason of possible environmental pollution at this candidate relocation site. I thought I would rather give up the plan if our longtime buddies in the fishery industry were really opposed us. Fortu-nately, both MLIT and our local government took proactive action at this time such as kicking off a committee to discuss the adoption of measures to prevent any environmental pollution, and the agreement by all fish processing industry was made at last. Looking back, I do think that we went through so many things and it was indeed a long journey.

To the Future for the Youth

I: Mr. Kidoura is 10 years younger than you. What is your view about him?

Y: I think he is doing a good job. He has a very good personality and is a highly respectable top management.

We have been in the same industry, and the two companies were located next to each other for a long time, and thus we have never been at odds with each other. Some shipowners may say, "I like Yoshida boats" and others may say, "I like Kidoura boats," but that's it. Certainly, there are some differences in terms of respective style of shipbuilding, but as long as the deal is done, I believe that we will eventually be integrated into one company.

I: Please tell me what is good about starting MIRAI SHIPS INC. and what challenges await the company.

Y: That is definitely that the shipbuilding industry in Kesennuma will continue towards the future. By this merger we became much more capable in terms of the number of ship types we will handle. In addition, thanks to the "shiplift," a highlight facility in the new shipyard, we are able to simultaneously work on 10 ships with around 100 tons of weight.

Also, thanks to numerous media coverage, MIRAI SHIPS INC. has become very widely known. If we recruit now, we will receive quite a few applications. This will be a great chance to pump young and fresh blood into the shipbuilding industry.

From now on, it is the young people that will lead MIRAI SHIPS INC., and it will be necessary to change the way we show them the ropes. Old-time craftsmen were told that "techniques are for stealing." Techniques were something to be leaned not by lecture but by your own observation at the spot of doing real work. But it does not go with young people anymore.

Shipbuilding is a tough business, but the harder you work, the more you will get in return. When you witness the completion of a ship you built, you will be surely rewarded by such sense that this is an indispensably fantastic job. I want to convey this kind of sense to the young generation and I want them to touch and feel it.

Anyway, the new company has got started, and we have no excuse. There is only one thing—just do it. The number of shipyards for building fishing boats are steadily decreasing in Japan, and Kesennuma is the last bastion. I am afraid that, if we failed, all fishing boats may be built overseas. Retaining the shipbuilding industry means protecting not only Kesennuma but also the entire shipbuilding industry in Japan. Personally, I have a sense of indebted to the people who worked to keep Yoshida Shipyard alive during my tenure, so I strongly believe that it will my obligation to meet the expectations from our shipowners.

2 | Takeyoshi Kidoura
Representative Director and CEO of MIRAI SHIPS INC.

Mr. Takeyoshi Kidoura, the president of MIRAI SHIPS INC., is 49 years old now. He is the youngest of the seven founders of the company. He has a diverse career, and also has experience of working in the United States in a completely different field of work. He had in mind to eventually be involved in work related to ships someday, but he was not particularly focusing solely on fishing boats. After studying yacht design, he returned to Japan and worked for a company in Shizuoka that designed and manufactured yachts for an international yacht race, the America's Cup. When he knew a few years later that Japan would not participate in America's Cup, he left the company. After many twists and turns in his career, he joined Kidoura Shipyard in 2008.

Serious Rivals

(Interviewer: I): You returned to Kesennuma in 2008, and took over the family business. Why did you decide to go back?

(Takeyoshi Kidoura: K): Toshio Kidoura, my father and ex-president of the company, developed an "electric propulsion" fishing boat together during his tenure in collaboration with Yanmar (a company manufacturing agricultural machinery, construction equipment, engines, and small ships). This was a fishing boat version of hybrid cars in which the screw propeller was driven not by a diesel engine but by electricity generated by an engine. This fishing boat was the first electric fishing boat in the world. Having witnessed this innovative achievement from outside, I thought that my family business company would have a work environment trying new themes, and I would be able to challenge new business there if I went back.

I: I think each of the four shipbuilding companies has its long history, but may I describe that you have been in the shipbuilding industry in Kesennuma without too much restricted by its old customs and habits?

K: That's right. I had been away from Kesennuma for a long time. During the time just after I joined, I simplemindedly imagined that all company would be very harmonized, which was not true. I was so surprised to see a very severe head-on competition at the time of bidding. I had imagined that there might be a well-organized horizontal relationship of the kind where at the time of bidding one would discuss with another beforehand, but there was not. In fact, there was not much communication with other companies.

I: Who was the president of Kidoura Shipyard at the time of the March 11 Great East Japan Earthquake?

K: It was Yuzo, Kidoura, my uncle. He was heading up activities of restoration for the shipbuilding industry after the quake, and I supported him from the working level. I guess he already had the intention of leaving the business to our younger generations soon.

I: So you were already making managerial decision at that point?

K: Yes, I was.

I: I heard that your uncle was the chairman of the Kesennuma Shipbuilding and Marine Industries Reconstruction Council which was established in July 2011. Had any discussions to build a new shipyard already started at that time?

K: Yes, it had. Within Kidoura Shipyard at that time, I was in a position to make various matters hearing people surrounding me, and my uncle confirmed my decisions accordingly. In April 2013, after two years from the quake, the Kesennuma Shipbuilding Complex Cooperative was established, and I became the representative director, though my title in Kidoura Shipyard remained a senior managing director.

I: You are still young, but Mr. Keigo Yoshida of Yoshida Shipyard is of the same older generation as your uncle. Wasn't it difficult for you to work with a person of an older generation?

K: On the contrary, it was good that I was young, I think. Because I was young, he listened to me about such things that tended to be somewhat confrontational otherwise as in the past. Mr. Yoshida clearly told me once that there would have been no merger if he would have had to talk to my uncle or my father.

Challenges Faced by Kidoura Shipyard

Takeyoshi Kidoura

I: Your uncle was the president at the time of the Great Earthquake, but your brother succeeded in 2012, and you replaced him two years later in 2014.

K: When my brother was the president, it was the worst time for Kidoura Shipyard in its long history due to the quake. Although there was a huge demand for reconstruction, it was very risky work. Shipowners wanted us to build fishing boats, saying they would lose much of their business if they did not leave the port by August, and if they did not leave the port, their crew would have to fall apart. If it happened, they say, they would be unable to get back on their feet again. Therefore they asked us to build the boats by all means.

Being witness to the difficult situation of the shipowners having business relations with us for generations, we could not able to decline their offer by making reason that we were also in a very tough situation. In addition to this emotional element, we also looked to supply chain issue where our decline could potentially lead to disappearance of fishing boats and fishermen in Kesennnuma. Our decision as a shipbuilding company was that we must accept the requests even if there are many foreseeable problems.

Subsequently, we arranged for workers who could move to our place from various locations all over the country in order to shorten the construction period. But we realized that even though we were compelled by circumstances to do so, arranging for people without sufficient skills did not necessarily lead to the shortening of the construction period. After all, things fell into a negative spiral and the more we worked, the more the situation worsened and there seemed no way out. The impact of this situation manifested during the time my brother was the president.

I persuaded my brother to step down and appoint me as the replacement because I believed that the company needed to drastically change its way of doing business to cope with both its business crisis and the recovery from the quake. I had overseas experience of starting a new business all over again from scratch, and I thought that I would be more suited to such a job that spins the wheel of planning and execution very quickly.

I: How did you rebuild the company?

K: The first action was to change our past style of relying on individual abilities. I changed it to a more systematic style in which all of board of directors share a common recognition about status quo, analyze its root causes, determine fundamental counter measures altogether and confidently cascade them to respective working level. Appropriate process control over the amount of work taken in the short term as well as increasing our capacity in the longer term were most critical because we had painful experiences after the quake which caused troubles to our customers because of our lack of process control together with insufficient capacity. It was easy to say so, but not easy to realize. Once those specific goals were set, our senior management leaders showed their leadership to their staff and craftsmen underneath. Thanks to their cooperation and efforts, outcomes began to emerge in about a year.

I: May I understand that you tried to put your company on a growth track by carefully looking to amount of work in the future and incorporate it into your business plan rather than dealing with today's work blindly as requested?

K: That's right. Looking to the timeline for not just six months or a year ahead but a little further, managers and employees shared the notion and tirelessly engaged with very plain tasks just like making muscles while slashing fat. Although we have been merged into MIRAI SHIPS INC., the process is still on going. The work brings our people in, and the people bring our work in. That is kind of a chicken-and-egg situation, and that is a very tough part of our business. It will be too late if you rush to get your people once orders are taken. We really needed to walk away from our old and reactive way of just fulfilling the gap.

I: Are you promoting rejuvenation in your workforce?

K: In our Kidoura factory, the average age of the employees just before the quake was late 50s and we had many employees close to their retirement age then. But today's average is about early 40s. As inexperienced young employees aged around 20 keep joining us, it will be critical how to pass down to them the techniques inherent to experienced workers.

But the speed of growth of our native local young employees is really eye-opening. They work more diligently than those who hop areas every time their work is done. Brushing up skills of those youngsters together with promoting mechanization and automatization will allow us to keep our technical level equally high as in days heavily dependent on experiences veterans.

Future Prospects of the Shipbuilding Industry

I: What kind of business development for the future do you have in your mind? Please tell us the prospects of the shipbuilding industry and the roadmap for the growth of MIRAI SHIPS INC.

K: Our role is to deliver and maintain boats needed by people involved in boat fishery and the fisheries industry. We noticed from our experience in the quake firsthand that the future of shipbuilding industry in a region are co-related with the future of boat fishery and the fisheries industry. MIRAI SHIPS INC. has no option but to grow.

I think there is a possibility that boat fishery will become increasingly bipolarized in the future and will be divided into coastal and inshore fishery with relatively small boats and offshore and pelagic fishery with large boats. As small and medium fishing boats are the core of our business portfolio, we are looking to maintain our business and to explore and offer new services that are different from those offered before.

What I am currently most interested are how the demand for large fishing boats will evolve and how we should be prepare for it. Foreign fishing boats, which may compete with domestic fishermen in the same fishing ground, have already increased in size and therefore, if the domestic shipowners want to build equally large fishing boats, we may end up with competing with foreign shipyards to get those demands. Our newly-launched shipyard is very capable of building such large fishing boats and will be able to meet our domestic customers. However, our newest facility only will not guarantee our future business of course. In order to ensure our business, we need to understand now how their future requirements for their ships would look like and incorporate them into our today's work so that we will be able to fully prepared when they come to us someday.

I: You mean that the merger of the four companies has increased your competitiveness as a shipbuilding company?

K: I think so. Our four companies may look similar at first glance, but our customers and types of fishing boats are quite different. Now that we can complement our experiences, know-how and information of each thanks to the merger and therefore I believe that we have become much more competitive than in the time of the quake. I am waiting for an environment to come where we will fully prove it.

I: How does the communication go within the merged four companies? What about with Mr. Yoshida in particular?

K: I think that not only Mr. Yoshida but also Mr. Sawada and Mr. Kosaba are all on the same page. Our factories were severely damaged by the quake, and we have to force our employees to work in a harsh working environment. We were confronted with the reality that we could not continue shipbuilding in the same way as before. We could not draw a vision of the future, or even worse, we could not see any path to full recovery.

If so, does it mean that we will put an end to shipbuilding in our generation? No way. We want to restore the shipbuilding industry in Kesennuma to the level of pre-quake at least, and then to entrust it to our next generation. We want to pass such a company down to the future where people can feel joy from their engagement in the shipbuilding industry. The deep thoughts are fully shared by all of us without expressed in words.

This is definitely our starting point and goal, and I think it does not matter if we are different in detailed arguments. It is obvious that we have differences derived from respective business and experience in the past, and I am convinced that the seven members will keep mutual respect and make one team best utilizing mutual experience in our future organization.

Shortly before the Start of the New Factory

I: The new factory is about to start operations. What thoughts and determinations do you likely to have on that day?

K: We are in the final stage of adjustments and preparations for the establishment of the new factory, and we will continue to make our every effort right until that day, until the very end. The merger of the four companies into one company is a major change for people involved in the shipbuilding industry in Kesennuma, and we can say that we have opened a new chapter in history. There is no doubt that our employees will have a hard time in getting used to the new environment, but it will not be something like walking away from work of shipbuilding and bringing up trees in the desert. What we are going to do is to make the most of our experience, to continue our shipbuilding business in a much better working environment and with a much better organization, and eventually to entrust it to the next generation. It is very simple and easy to imagine. I believe that, as long as we can go back to the point, we can keep committed to our work with sense of colleagues at MIRAI SHIPS INC. even if each used to belong to different company or factory before. We have long cultivated the field and sowed the seeds, and the tiny bud, that is the MIRAI SHIPS INC., has finally sprouted from the ground. We are a team of members who will bring it up altogether. If everybody believes so, the integration will proceed very smoothly and quickly, I believe.

I: Is it you, Mr. Kidoura, who will lead MIRAI SHIPS INC. as the president? Or will you lead it together with the heads of each factory on a collaborative basis?

K: Either way, and there could be other ways. The most important thing is ensuring that the company will continue, and I will be happy to fulfill the role required flexibly in accordance to surrounding situation and the environment, just like water that changes its shape according to the shape of the vessel.

I: You mean that your role will gradually emerge in the course of interactions with your management team, right?

K: I believe so. My task as the president at the time of the launch is to set the business on track and to pay back the bank loans as scheduled. It will not necessarily be required that I will pull everyone forcibly, and I do not think that we should make our organization dependent on a prominent leader with qualities and abilities. I'm looking forward to further discussions with everyone regarding what type of organization would be desirable to achieve both stability and growth.

I: Will you tell me what the source of your motivation is, Mr. Kidoura?

K: No doubt I like the job and people of building ships and it is in my formative experiences. I once pursued a career of building yachts instead of fishing boats, but I returned to my home town to work for my family business. Then the quake occurred, and it has brought MIRAI SHIPS INC. If things go well, the new shipbuilding company could become the one with the state-of-the-art people and technology so that someone come to us to order racing yachts. My dream will come true then and moreover, as a good precedent or benchmark, the company may draw attentions from more people, which could lead to achieve much bigger dreams of someone.

I believe in a following phrase that "If you have passion, conviction, wisdom, and comrades, most things can be accomplished." This is the words of a business person that I have greatly respected, and I want to use the energy from these words to achieve my mission of restoration and reconstruction. From now on, my desire to be an executor and a supporter in making dreams and hopes realized by way of work for shipbuilding will keep driving me move forward.

The MIRAI SHIPS INC. plan started amidst the post-tsunami, mud-scraping work, and became a vehicle conveying hopes and aspirations of a very small number of people involved. In the beginning many people advised us directly or indirectly that it would be impossible or nearly impossible to make it happen considering the history and the situation at that time. But with passion, conviction, wisdom, and our many comrades, the new company and facility were completed in 2019. I would like to take this opportunity to convey my appreciation to all people concerned. I promise that all of us, both management and employees, will strive to make this company the *mirai* to many people as written in our company name. The continued support from everyone will be highly appreciated.

3 Tsutomu Sawada
Director of MIRAI SHIPS INC.

Mr. Tsutomu Sawada, former president of Sawada Shipyard, says that he is most pleased that the working environment will be greatly improved by the completion of the new shipyard. But he says very straightforwardly that the old style of working will not work if we are to attract young people to the shipbuilding industry. He feels the changes of the times first hand. On the other hand, he also says that he wants the new shipyard kind and respectful to the elderly employees. New equipment will ease their decline in physical strength, which will consequently allow techniques and experience of the elders passed down to young people. Knowing well the basic circumstances, Mr. Sawada talked about challenges MIRAI SHIPS INC. will face.

The Sea of Darkness

(Interviewer: I): Tell me how Sawada Shipyard was damaged by the tsunami?

(Tsutomu Sawada: S): All our three berths were engulfed by the tsunami, two boats under repair fell from the berth, and one was washed away by the tsunami. All these boats had been entrusted to us by our customers.

I: Was it covered by insurance?

S: We have the Ship Repairers' Liability Insurance, but tsunami damage tsunami was not covered but our customers claimed the damages to us. None of us did not know how to do it after all because it was a natural disaster. We were just so relieved that there were, fortunately, no casualties among the factory employees or among our relatives.

I: Were you able to get into the subsequent restoration work smoothly?

S: Due to land subsidence, the land of our factory gets submerged during high tide. So we laid rails and raised the land. For constructing these rails we bought used materials by ourselves, and also paid for the labor cost. Those costs were ultimately covered by the Group Subsidy though, but it was really tough right after the quake because we had to manage everything by our money at hand.

I: So you had a hard time in the initial stage of immediate restoration?

S: Some companies were soon provided with the subsidy speeded up their restoration, but other companies like us were not able to look ahead clearly because of delayed decision regarding subsidy, which made us very anxious anyway. It was like being thrown out into the dark sea without sight of a lighthouse.

Tsutomu Sawada

Jumping off the Tokyo Sky Tree

I: I heard that it was difficult to reconstruct your factory at its original location due to the land subsidence. Did you want to restart by the relocation from the beginning?

S: I did not deny the relocation. But I knew that it would take time, and I partially wanted to stay and continue my operation at the same place. It was not that simple. If I think immediate restoration of more important, then the choice would have been staying there by repairing damaged equipment and so forth.

But in hindsight, it was necessary and right to restart by making a new company. Improvement of our working environment was an inevitable path for the sake of the future of the shipbuilding industry in Kesennuma. A hard, dirty, and dangerous workplace is totally unacceptable today, don't you think? The younger generations will not join us and our shipbuilding industry will decay. The future of the shipbuilding industry will never be bright unless we make an environment that is conducive for young people who want to build ships or process iron.

I: The new shipyard is about to start operations. May I say that all the things have finally been done for this huge project, and you have unloaded your heavy luggage?

S: No, it is far from it. From now on, we are starting business with sales scale larger by one digit from our time. Our work is going to step up to a different level from the ones that respective company used to do. We cannot take it easy saying "It will work out anyway." Basically, we have been taught by our predecessors that "Do your business within your affordable limits." and that has been exactly the secret to continue doing business for a long time. But what we decided this time demanded unprecedented preparedness and determination on our respective end. It felt like a leap in the dark, like jumping off from the top of Tokyo Sky Tree Tower (smiles).

I: Are you saying that you have lots of concerns?

S: Yes, but it's not just us. Our customers are also concerned about it. They ask us, "Will our boats still be really taken care of from now on?" They are very long-time customers of ours. Even if I explain to them that we will continue to work for them in the same way as in the past, they remain skeptical by saying "Is it really the case?" I guess other three companies may have similar situation.

In the new company, we must keep our attitude of meeting customers' needs with agility. I would like to tell our young people to inherit and value those customer relationships which has been cultivated and succeeded over a long period of time.

The Shipbuilding Industry in Kesennuma to Our Future

I: I heard that the fishing industry in Japan is generally shrinking and the situation surrounding the shipbuilding industry is also quite tough.

S: In the past, our business enjoyed better times for a long period than it does today. There was a time when salmon and trout in the North Sea were abundant and popular, and then there came a period when tuna reigned. Now we deal with various kinds of fish like tuna, bonito, and saury, which means that the shipbuilding industry in Kesennuma still has its reserved strength to manage this diversity, and things are not so bad after all.

For example, Kosaba and Sawada are competitors, but each of them has a fixed customer base. They will share work when they are busy, but they will never steal work from each other. Kidoura and Yoshida, who are both new boats builders, used to be rivals maybe, but they have their own distinct customer bases now.

I: It sounds that the new company will likely demonstrate its collective strength of respective merged company. What is your short-term goal of the new company?

S: The first thing is to maintain its status quo. We have to acquire proficiency in our newly introduced machines and equipment which we have never used before. It will be important to demonstrate our capabilities to the full. Avoid any accidents at all costs. To make our operation sustainable, we will work with our greatest caution and care, with keeping in mind that there is no room allowed for accidents or injuries.

The other day, when we were inspecting a bonito boat the other day, a wire snapped, and the boat was suspended on one side. We had to insert a wire, but there is nobody in Kesennuma who has the technique to do so. Everyone has got old. Then a trawl fishery customer happened to see it, and took care of it saying "Okay, I'll do it." I thought he was a godsend (laughs).

As such, there have been times when customers help us in various ways. We often say, "You and I are on the same boat when in trouble." Such relationships with customers are our asset.

I: It would be good if such people-to-people trusted relations will be inherited by the new company.

S: That's right. We were taught and raised by older generations indeed. In our new factory are highly automated and ships will be lifted by pushing one button, which will be physically helpful for our elderly workers. I hope this company will be a place where elderly employees can work in piece so that they can teach various techniques and pass them to younger generations. I want the younger generations to further refine the succeeded techniques in their job. That is my ideal.

4 Yoshinori Yoshida
Director of MIRAI SHIPS INC.

Mr. Yoshinori Yoshida is the fourth-generation president of Kosaba Shipyard. When he joined the company in the 1980's, fishery with ocean tuna boats registered in the Kesennuma registry was in its heyday, and boats got larger and larger. In recent years, however, the number of boats declined. The quake and tsunami struck in the midst of decline. The land subsidence also caused serious damage to the industry hard, but Mr. Yoshida rather regarded this as the first and last opportunity for revitalization of the shipbuilding industry in Kesennuma, and was determined to work aggressively on the project of MIRAI SHIPS INC. He aims to achieve further growth in the shipbuilding industry in Kesennuma through cooperation and harmonization of the companies.

A Long-Standing Cooperation System

(Interviewer: I): How was the damage caused by the quake in the case of Kosaba Shipyard?

(Yoshinori Yoshida: Y): We replaced all the bent rails with new ones and ensured that at least one of the berths could become operational. It was around November in that year when we were finally able to restart our work. It took time. In addition, the coastline shifted about 10 meters inland due to land subsidence and our work area shrank. For large boats, work had to be done while being immersed in seawater. The same thing went with other shipbuilding companies.

I: I heard that members of the merged four companies are known to each other since old days.

Y: Yes, we have known to each as we belonged to the Kesennuma Shipbuilding Cooperative which had existed since before the merger. But things were not that easy, as each company has a long history, and we were also competitors, in a way. I heard that, our predecessors were in a kind of collaborative mode so that shipbuilding companies would vitalize

Kesennuma and it seemed that they tried a plan to save money to buy land jointly, but it ended up with a halt, and did not reach a discussion of some merger. There had been such explorations for a long time, the severe and extensive damage caused by the quake did prompt us to seek the path seriously.

My Uncle's Encouraging Words

Yoshinori Yoshida

I: I heard that 18 companies formed the Kesennuma Shipbuilding Complex Cooperative and declared to build a new shipyard two years after the quake. Please tell us about the process from around this point to the merger.

Y: Eighteen companies related to the shipbuilding industry gathered together and formed the Cooperative because it was necessary to do so in order to apply for subsidies. However, as the plan was gradually taking concrete shape, realistic issues came up as to who would arrange and pay back for expected financial burdens and who will take a role of a guarantor, and so forth. In the end, the seven companies made by four shipyards and three other companies stood up to become responsible for the project.

Although there were many twists and turns, we moved forward with determination for the merger, thanks to the efforts of Kesennuma City Hall, and the contributions by MLIT and Deloitte Tohmatsu. The prospect of receiving public funding (the Reconstruction Support Project Subsidy for the Shipbuilding Industry) for the construction of a new shipyard, was the most significant factor when I look back now, and it was a big relief for me.

I am thankful that the City Hall mediated for the merger plan of the four companies, and that Mr. Yoshikawa of Deloitte Tohmatsu and his team provided specific advice when we did not know what to do and how to go about the merger.

Also, when we needed to calculate a merger ratio, the four companies had an opportunity to share their respective financial data for the first time. Until then, we did not even know the exact size of sales and profit of each company. I guess that we got a sense of positive mood thereafter if we, the four companies, might be able to make it somehow altogether.

Additionally, I was personally greatly influenced by what my uncle in Shiogama told me. He really cared about my situation after the quake, and told me that "If Kesennuma is ready to pay, you had better think of working on the new thing as the first and last opportunity in your life. It will be now to choose the merger if it is your option." His words set me thinking more seriously about the merger.

I made up my mind that the most important thing would be not to protect the name of my company but to continue our shipbuilding business with a history of more than 100 years in this place. I discussed this with the other presidents and we reached a conclusion—it would be quite difficult to continue the shipbuilding business in the future if we keep the same way as in the past, but we might be able to make it somehow if we get together and make one company.

I: I heard that you became considerably sick during this time in fact. Wasn't it quite difficult for you to work on the merger plan while continuing restoration work after the quake?

Y: Yes, it was. I suffered cerebral infarction in 2015 and it took me nearly two years to be back to normal. My doctor in charge then advised me, "Reduce your workload. Entrust your workload and your roles and responsibilities to your employees as much as possible." I thought he was right. So, I began to delegate my work as much as possible. My management team and employees all worked really hard and I had nothing in particular to worry about during this period.

Connecting Technology and Feelings to the Future

I: The new factory of MIRAI SHIPS INC. will be able to build and repair larger ships than before once it gets started with full-scale operations. I heard that it will be able to deal with as many as 10 ships simultaneously by the introduction of the shiplift system. Do you have any ideas or expectations towards possible expansion of your business?

Y: Yes, it will be good if our scale of business expands, but the most important thing is that the employees of the four companies will be able to be united and work together as members of MIRAI SHIPS INC. as soon as possible. Since the four companies with their own respective technologies will get together, it would be a shame if we work in disunited way and don't hone our skills together.

If you categorize the four companies, Kidoura and Yoshida specialize in construction and repair of new boats, and Kosaba and we, Sawada, specialize in repair. And we all are different in terms of types of boats to deal with, materials to use, and process to work with. We, Sawada, don't know much about new ships, but instead know well about woodworking technology. As the technology can be used to building new ships, I think we can contribute to the new company in terms of reducing production time and increasing work efficiency.

Before the merger, our main discussion points were on how to reconstruct the shipbuilding industry in Kesennuma and whether a merger could be done. However, as long as the merger is done, it is important to consider to what extent we can draw benefits out from the merger.

We must think of fostering interaction among our employees and must decide job description in our new organization before the start of operations. It is imperative to ensure that employees would do their job properly once it gets started.

At this stage, some interactions have already begun as we sometimes undertake work from Kidoura. But when it comes down to how to process those work, we still discuss it only internally, and it does to come to a point where we say that "we are doing job together with Kidoura." It is partly because we are extremely busy these days with our current work already undertaken and therefore there is no time to think about new way of doing.

Of course, I don't think that things will go well by merely setting rules in advance. Once the new factory gets started and everyone gets started to work in one building, then real and concrete ideas about new ways of working will come up and subsequent improvements will be made.

I: Any thought to the name of Kosaba Shipyard which will disappear?

Y: It is very sad that the name will disappear during my tenure, but I think it should be fine as long as we can retain our techniques and ideas associated with shipbuilding which have been inherited so far. I would like to pass down our precision handiwork inherited from our predecessors from the era of wooden boats to the next generation. I believe it is important to continue the legacy of the shipbuilding industry in Kesennuma and to relay it to the future. I think our thoughts are engraved in the name of "MIRAI SHIPS INC."

5 | Masayuki Fujiki
Outside Director of MIRAI SHIPS INC.

Mr. Masayuki Fujiki leads the shipbuilding company "K-Yard" which special-
izes in the construction of aluminum fishing boats. He is originally from Kes-
ennuma, and the company also used to operate in Kesennuma before the quake.
However, both his house and factory were swept away by the tsunami, and he
was compelled to give up the idea of restarting the factory. As scallop boats built
by the company had been much favored by shipowners in Hokkaido, and Mr.
Fujiki found an appropriate factory for relocation in Nemuro City, he decided
to move to Hokkaido. However, his feelings for Kesennuma, where he was born
and raised, only increased after the relocation. When he heard of the launch of
MIRAI SHIPS INC., he made up his mind to participate in the project, for the
sake of reconstruction Kesennuma, and the return to his hometown.

From Manufacturing Aluminum Out-fittings to Building Aluminum Boats

(Interviewer: I): Please tell us about the history of K-Yard.

(Masayuki Fujiki: F): The predecessor of K-Yard is Kishu Kogyo, which was
founded in 1947 by my father, Tetsuro Fujiki. The company originally
built the "hull" of steel boats (the outer part of the hull) and delivered our
products to shipbuilding companies. We undertook the whole process of
design, ironwork assembly, welding, and plate bending all together, and
helped small and medium shipyards not only within Miyagi Prefecture
but also in other regions such as Mie and Aichi Prefectures. We also
delivered our products to Kidoura Shipyard and Yoshida Shipyard.

In around 1980 and thereafter, the work on aluminum out-fittings
and superstructures began to increase, so we established a subsidiary, Kiyo
Kogyo, which handled aluminum processing. Furthermore, we began to
directly receive orders for construction of aluminum boats starting from
1985. The first ship we built was a seven-ton high-speed leisure fishing
boat.

I: When was the present K-Yard formed?

F: It was in 2003 when two factories were relocated to Namiita Factory and consolidated. In 2016, the number of aluminum boats we had handled exceeded hundred. We handled not only fishing boats but also catamaran type passenger boats which toured around Matsushima, a sightseeing spot. We received orders not only from Miyagi Prefecture but also from a wider area including Iwate and Aomori Prefectures.

When orders decreased, I often went to Hokkaido to get business. I think it was before the beginning of the 2000's when orders from shipowners in Hokkaido began to account for the majority. Scallop boats have a good reputation in particular, and we have already built about 60 boats to meet orders from a wider area including Abashiri and Wakkanai in Hokkaido. Our business after moving to Hokkaido is going fine thanks to stable demand of after-sales service for our old boats and continued orders for of new boats.

I: What are the characteristics of aluminum boats built in your company?

F: I think that it was very good that we pursued our original style of hull form development as well as construction technologies as much as possible by introducing 2D and 3D computer-aided design (CAD) very early when small shipyards did not have even a personal computer. As a result, our boats came to be famous by unique hull forms with multiple curved surfaces and enjoy good reputation from shipowners in terms of high speed and better fuel efficiency.

Decision to Relocate after the Disaster

I: Why did you choose to move to Hokkaido?

F: Before the quake, the president of Kudo Shipbuilding Ironworks in Nemuro, a great friend of mine, offered me to move over there and utilize her factory and equipment. She had succeeded her business from her husband who had passed away, but wanted to retire for her age reason. But I could not make an instant decision to her offer because I was based in Kesennuma. When the quake struck in 2011, I was in Nemuro for business and returned to Kesennuma in the morning of the fourth day after the disaster. All of my family members and company employees were

safe, but my home had been washed away, and
the factory had become completely unusable
due to land subsidence.

To be honest, I thought that it would be
unlikely to make a recovery in my ability.
Meanwhile, we were working on a scallop
boat ordered and already paid by a shipowner
in Abashiri, and I thought that I would need
to complete the work. I asked the president
of Kudo Shipbuilding Ironworks for help
and she kindly accepted it so that I could uti-

Masanori Fujiki

lize her facility. This was the start that we used their factory facility and
machinery at Hanasaki Port in Nemuro City.

Our damaged factory in Kesennuma got cleaned up in about a week
by our employees with help of volunteer people. I traveled back and forth
between Nemuro and Kesennuma after then and frequent transportation
by truck costed a lot of money.

I: Did your employees at K-Yard move straight to Nemuro?

F: No. Each had his or her own personal circumstances, and some of them
could not leave Kesennuma. In the end, four out of 12 employees chose
to remain in Kesennnuma, and we moved to Nemuro together with the
eight employees. Some of subcontractors who had supported us came
with us.

In May, two months after the disaster, we restarted construction of the
half-built boat, and in December of that year, five months behind sched-
ule, our first reconstruction boat was completed. We managed to deliver
it to the shipowner. As I sensed that I forced my employees to come with
me, I was always telling them that they can go home anytime at my cost,
and I encouraged them to return home at least three times a year. I l also
had a big anxiety about moving to Nemuro under very uncertain situation after the quake, but I had only one thing in my mind—make money
and pay salaries to my people—and just devoted myself to it.

I gradually hired local people in Nemoro and now have ten young
employees who grew up just fine. Currently, all employees in our Hanasaki

Port are all native to Nemuro. They have already built two ships single-handedly, which are as good as those built previously.

Revive Kesennuma by All Means

I: In the meantime, in the time that the four shipbuilding companies were heading towards the merger, what was your view about the movement?

F: The quake washed away my factory and, to be honest, I did not know how to set my direction at the time, and did not have any alternative but to devote myself in my daily work just to survive.

Although both Mr. Kidoura and Mr. Yoshida insisted that the merger would be the only way, those future prospects did not click me at all at that time, and my feeling was more like, "I will leave everything to you guys." As I had already left Kesennuma and was working in Nemuro, I wondered if I was allowed to part of the project and also a bit skeptical about whether it would really materialize or not. Although discussions kept going in Kesennuma, I was not able to attend them and did not know much about details.

However, my colleagues and comrades reached out to me in Nemuro and kept me in the loop by saying, "Let's do it together," which made me very happy.

I: Did you ever have the desire to return to Kesennuma one day?

F: Yes, I did want to "come back and restore my business by all means" partly for my employees who remained in Kesennuma. I am genuinely native to Kesennuma and the registered address of K-Yard headquarters still remain my home address in Kesennuma. However, it is still impossible even to rebuild my home due to problems related to tide embankment. In these circumstances I cannot go back home even if I hope to.

Promoting Diversification of MIRAI SHIPS INC. with Aluminum Boats

I: MIRAI SHIPS INC. will start full-scale operations in April 2019. In what way are you planning to be involved in it?

F: The new company would have been good enough by merging just four of the shipbuilding companies, but they called us, an aluminum boats builder, to join project. Our company will fulfill the expected responsibility with related to aluminum boats.

In recent years, fiberglass reinforced plastic (FRP) fishing boats are becoming larger, and FRP fishing boats of over 100 tons have become the mainstream for inshore bonito boats off the coast of Sanriku. Similarly, aluminum boats are also expected to get larger. I would like to build a large bonito-boat-sized aluminum boat someday, which has been my dream since a very long time.

When the new factory gets busy both in new boats construction and old boat repair, they can manage it by bringing some of them to our factory in Nemuro. It will be necessary to accommodate these boats flexibly so that both facilities can avoid idling. That is my hope.

Before we started working on aluminum boats, we were building the hull for steel boats as I mentioned, so I know how to deal with iron. Also, as we were a supplier to both Kidoura Shipyard and Yoshida Shipyard, we are aware of their way of doing and, I hope that we could help MIRAI SHIPS INC. in such domains that they are not familiar with, but at the moment we will see.

Mr. Keigo Yoshida and I are of the same age. I was born in March and am one grade ahead of him. Keigo was born in May, and Mr. Tsutomu Sawada of Sawada Shipyard was born in February. All three of us (*san-nin*) who were born in the year of the Rooster (*tori*) are called "Suntory (*san-tori*)" (a famous beverage company) (laughs). Mr. Yoshida says, "We need you to build more aluminum boats. Let's do so in the new company." My company used to deliver the hull of steel boats to Yoshida Shipyard. Also, when K-Yard moved to Nemuro, Mr. Yoshida accepted our employees who could not leave Kesennuma. As the MIRAI SHIPS INC. project already existed at that time, I told them to work together again when we get together.

About five years ago, we built a new aluminum boat on the berth at Yoshida Shipyard as a kind of trial. At that time, our former employees who were re-employed by Mr. Yoshida joined us and together we built the ship, and unveiled it to Mr. Kidoura and Mr. Sawada.

I: You have a base in Nemuro. How do you manage your work in the future?

F: I will go back and forth between Nemuro and Kesennuma for the time being, but I am going to be stationed in Kesennuma in about three years later. I am currently training candidates for leaders at our Hanasaki Port Factory in Nemuro so that I can stay there only for about one week or so every month. As we have good tools such as e-mail, so there will be no problem. In the future, I will do more sales activities by setting a base in Sapporo.

Ensuring the Youth Experiences the Sense of Achievement That Accompanies the Construction of Each New Ship

I: How do you train employees of MIRAI SHIPS INC.?

F: I want them to learn a wider range of shipbuilding technology for both aluminum boats and steel boats. For example, if there is a surplus in people for building a steel boat, then some of them can be shifted to help building of aluminum boats. I want to nurture such personnel who can perform in multiple tasks. I also want to further brush up technology for building large aluminum boats because it will be difficult to make profits by only building small boats.

I: Does technology for building aluminum boats differ from the one for steel boats?

F: As long as shipbuilding technology concerned, steel is more difficult to deal with. But both steel and aluminum are metal, the design method are quite similar The properties of steel sheets such as hardness vary depending on manufacturers such as Kawasaki Iron Works and Nippon Steel and so forth, but on the other hand properties of aluminum materials do not. And also, aluminum boats often contain structural materials such as ribs, you will need various technologies such as cutting out many parts and bending them precisely as designed.

I: Are you attracted by working for shipbuilding? Why is that?

F: I've been involved in this job for over 40 years, and I think it is a really profound and fascinating job. There is no other job where you can learn wide variety of things and can get sense of accomplishment everytime you build a ship, all of which makes you so happy. Moreover, you will get a tremendous sense of unity and camaraderie that emerges from your work. I believe that the role of MIRAI SHIPS INC. is to convey this thrill, this sense of achievement to junior fellows, to nurture more people in the younger generations, and to connect this to the development of the shipbuilding industry. I cannot think of my retirement until I witness how the future of MIRAI SHIPS INC. evolves.

When I entered this industry, we were in the generation immediately after the baby-boomers' and there were very few people in between. Currently, the number of people in our generations is decreasing and the main body at K-Yard in Nemuro is the people in thirties. The same thing goes with shipbuilding industry both in Tohoku and Hokkaido and there is shortage in people who will educate their successors to lead the next era.

At MIRAI SHIPS INC., I think that we senior management should not set back in respective role of department heads but instead should take our proactive actions to raise up such people who will take on these roles in the future.

For this purpose, it will be naturally necessary that we welcome and train young staff. The shipbuilding industry has been deemed as a hard, dirty, and dangerous workplace, but our MIRAI SHIPS INC. has brought a chance to dispel this prejudice. Fortunately, the new company is drawing big attention. It has the latest equipment such as a shiplift, and various aspects have been taken into consideration in terms of the working environment.

Shipbuilding is the job that we can definitely be proud of. I want more young people to join this industry. And, I want to make Kesennuma the center of building fishing boats in Japan and want to see it become a vibrant city resounding with the cheers of young people. I am looking forward to the coming April when MIRAI SHIPS INC. starts its operations.

6

Yuto Ishikawa
Outside Director of MIRAI SHIPS INC.

Mr. Yuto Ishikawa, the chairman of Ishikawadenso, is the oldest among the board members of MIRAI SHIPS INC. He had just let his eldest son, Takayuki, take over his position as president in January of this year. Apart from his main business, he is known as a social philanthropist. On the day of the quake, he was working on putting waste cooking oil collected from school lunch centers and restaurants in the city into drum cans in the vicinity of the company for the purpose of recycling waste oil as fuel. Mr. Ishikawa, who is known as a theorist among the seven business owners, asserted that building a new shipyard was the only way to pioneer a new future in Kesennuma.

Restarting from Nothing

(Interviewer: A): I believe that Ishikawadenso is primarily engaged in business related to boats. How did you manage work right after the quake?

(Yuto Ishikawa: I): Before the quake, electric work for boats accounted for nearly 80 percent of our business. We were also doing non-boat related electric work in parallel, and I was hoping to increase it up to half of our business but then the quake struck us. As local shipping-related companies were severely damaged, I became very anxious about our future. One day in April, the chairman of the Kesennuma Chamber of Commerce and Industry called in President Kidoura (Mr. Yuzo Kidoura, uncle of Mr. Takeyoshi Kidoura, the president at the time), Mr. Onodera and me and told us to make each of the shipyard, motors and electric equipment temporarily restored and ready so that we would be able to welcome the bonito boats arriving at Kesennuma Port to land fish in June. What an unreasonable request! (laughs)

A: But electricity and water supply had remained disabled at that time, right? How did you get this work done?

I: We could not even restore our equipment damaged by seawater because there was no fresh water. There was also no electricity nor the people to work. Some of our employees had to be in evacuation centers. But motors were brought in to our factory one after another from the shipyards. So, I went to Yokohama to get a generator from a friend of mine who was willing to lend. When it came to water supply, another friend of mine in Iwate Prefecture brought in a 600-liter plastic container and a motor pump to draw water from the river. As I had

Yuto Ishikawa

more and more employees participating in the restoration work with simple lunches like rice balls and a water bottle, I gradually increased our workload.

A: Then you became able to deal with increasing requests?

I: No, not really. An enormous amount of motors was brought in to us for emergency repair. Besides, motors immersed in seawater needed to be boiled for 24 hours in fresh water to remove the salt. That took us too much time, so we could not deal with the brought amount at all. I asked my acquainted chemical cleaning agent manufacturer for an advice and tried a method to wash away salt in a short time. We barely managed to get through. After all things got done, then I noticed if it is ever paid for, because all of those our customers had suffered as well (bitter smile). Ultimately though, all the payments were honored.

Road to Restoration

A: When did you start seeing prospects for your business?

I: When the busy season of bonito boats was over and summer arrived, we began to be asked to replace the fish lamps in large saury boats, with LED lights. Filament lamp bulbs had been used before the Great Earthquake, but these consumed a lot of power. The generators had been swept away. If these fish lamps could be replaced with LED lights, large generators would not be needed anymore. And the subsidies were applicable to the equipment. Although we gave discounted prices to those customers who had also suffered extensive damage, it ended up with a special demand derived from the quake for us as electric equipment providers. This made me feel relieved that I could manage to continue my business without firing my employees.

As we had very little work right after the quake, I had to ask them to endure salary cuts in about 30 percent. Instead, I told them, that they can come to work at any time as they like. It was okay if they can come only at 10 in the morning, or if they want to return home at 2 in the afternoon. Finally I was able to restore their salary level back to the original one by autumn without firing anyone.

A: After going through such hardship, did your business get back on track?

I: No, things remained tough after that as well. Before the quake, there were several electric equipment companies, and there were three including us who dealt with electrical installation for new boats. The three companies were competing with each other, but in the quake the two out of the three had either their entire building washed away or their main workforce perished, and we were left as the only company who could continue the operation. Boats had to be built anyway for shipbuilding companies to move forward, and we were asked to run with them by all means when they build new ships. We took the challenge and worked very hard.

A: Meanwhile, you have been working for the sake of industry since then.

I: Mr. Onodera headed the Kesennuma Reconstruction Committee for Fishing Port Functions, which was set up to get the Group Subsidy from the Small and Medium Enterprise Agency[13], and Mr. Yuzo Kidoura took his role of coordinating the whole industry. I supported both of them. Around the same time, it came up that the subsidy might be provided by the support of the Nippon Foundation, and we kicked off the Kesennuma Shipbuilding and Shipbuilding-related Industries Business Council. I was chosen as the chairman of the new organization and dealt with

---

13 The official name of the Group Subsidy of the Small and Medium Enterprise Agency is the "Restoration and Maintenance Project Subsidy for Facilities of Small and Medium Enterprise Groups." See Chapter One.

various coordination works with the Nippon Foundation.

The New Shipyards Had to Be Built

A: Mr. Ishikawa, you have always been involved in restoration movements that eventually led to the MIRAI SHIPS INC. Please tell us your thoughts behind it.

I: After the quake, we were left alone as the only companies that were able to handle electrical installation work for new boats. That meant, no boat will not be built in Kesennuma without us, which naturally gave us including employees a sense of mission. We could not but think of the future of Kesennuma.

A: Will you tell us the characteristics of the shipbuilding industry in Kesennuma?

I: There are less than 10 famous shipyards building fishing boats in Japan, such as Miho Shipyard in Shimizu City (Shizuoka Prefecture) and Niigata Shipbuilding & Repair in Niigata City (Niigata Prefecture). Both shipyards can take collective orders from shipowners thanks to their inhouse electricity and engines divisions. They will outsource when necessary but they undertake work comprehensively in one order.

On the other hand, in Kesennuma, shipyards, electricity, refrigerators, engines and painting are separately undertaken and it is a long history. For example, when a tuna boat returns to Kesennuma Port for repairs after one-year sail, one shipyard inspects and repairs only the hull, and the shipowner has to negotiate with other companies directly regarding the rest of work. Thus, all companies have to cooperate with each other, otherwise it would not be possible to repair just one boat. In another word, we are a part of shipbuilding company, and this is why our role is so important.

A: When you thought of the four shipbuilding companies from a longtime-partner viewpoint, you thought that it would be necessary to revitalize all of them by making them into one company if they really intended to overcome the crisis.

I: Yes, I did. That was the only way. There was no other option for all the four shipbuilding companies to survive because they had aging employees and facilities.

A: This time, you have introduced the shiplift system in which one can move ships up and down like an elevator. I heard that it will be the third such facility in Japan.

I: To be honest, it was extremely difficult to build and repair large, state-of-the-art fishing boats in our conventional slipway-type facilities. Accidents occurred once in a few years, with boats that had been dragged collapsing sideways. It was very dangerous to lift up a boat in an old facility beyond its capacity, and was a difficult task that only skilled workers could do together.

That's why, a work environment, where even young people could lift boats up and down more easily, was really needed. Of course, we have high expectations that the new facility will reduce the risk of tsunami damage. These aspects will finally be realized by the introduction of the shiplift system.

Why Kesennuma Has the Best Repairing Capabilities in Japan

A: What do you expect MIRAI SHIPS INC. to be evolved into in the future?

I: From now on, MIRAI SHIPS INC. will take all orders of repair collectively, and distribute the work to our partner companies, just like shipyards in Shimizu and Niigata do. Our business process will be changed. It would not be called as a top class shipyard if we get paid just for mediating but repeat the same things as in the past. We must carry out proper process control and quality assurance before delivering a boat to the customers, and will need to raise up our people armed with high skills.

Meanwhile, partner companies must strive to be thoroughly reliable as well. As they are in a subcontractor position, it is natural that vertical power relationships will be there, but I think that they can keep a sense of equal partners with us. They will surely get in trouble if the shipyard goes out of business, but conversely, the shipyard will lose to the other docks unless all collaborating companies keep healthy and robust. I wish the long-standing tradition and heritage in Kesennuma will prove viable.

A: I heard that the repairing capability of shipyards in Kesennuma has been high.

I: Yes, it is. I think the capability to repair large fishing boats, in particular, is the best in Japan. As ocean tuna boats do not return to the coast for a year, it is fatal if they get caught in trouble in the middle of the sea. Even when shipowners initially place only small-scale orders, craftsmen in Kesennuma work hard during inspection of boats to identify if there are any problems. They are committed to take all possible measures and make doubly sure that, "Nothing is left unrepaired" and that, "All repairs are absolutely essential." That's why everyone feels safe and secure when they dock in Kesennuma. I hope MIRAI SHIPS INC. will take the initiative and maintain this tradition.

Nurturing Young Personnel and Encouraging Early Inheritance of Leadership Roles

A: You will invest in MIRAI SHIPS INC. by yourself and take on your managerial responsibilities as a board member, correct?

I: It would be better to appoint motivated and meritorious individuals from the four shipbuilding companies. It is also very natural to promote such young individuals to the position of board members. Therefore, I myself would like to let a real person with spirit of shipbuilder take over my position as early as possible and serve as a technical adviser for electrical equipment if needed. Things will move smoothly as such even if relationships may be strained initially.

A: Nevertheless, I guess there should be such things that you would like to give your suggestions to the new management people as well.

I: Well, I still hesitate to make suggestions to ex-president of shipyards who used to be my customers (laughs). However, I do put forth my suggestions when I believe I have something important to say.

A: In Ishikawadenso, you let your son take on your position of president. How does he see MIRAI SHIPS INC.?

I: Of course, he is willing actively related to it. Working with the new company remains his main business. But he belongs to a different generation from mine, and is different in the ways of doing business from mine. I want him to make his contributions in his own style and capabilities.

In any case, I want younger generations to take over our roles early and to broaden the future horizons in Kesennuma.

7 | Shigemasa Onodera
Outside Director of MIRAI SHIPS INC.

Mr. Shigemasa Onodera, the president of Onodera Iron Works, is 53 years old now. After graduating from university, he worked for a ship engine manufacturer in Tokyo. At the age of 25, he returned to his hometown to take over his family business of inspection and maintenance work on marine engines and equipment. Immediately after the quake, he actively started moving toward the reconstruction of the shipbuilding industry in Kesennuma. He tackled with making his business plan in the "Future Management School" along with Mr. Takeyoshi Kidoura, who was the senior managing director of Kidoura Shipyard at that time. This plan became a blueprint of the MIRAI SHIPS INC. project. To the question why he got enthusiastic about this project, he answered that it was needed for the reconstruction of Kesennuma.

Starting from Mud-Scraping

(Interviewer: I): First of all, please tell us about your recovery initiatives right after the Great Earthquake.

(Interviewer: O): Before the quake, we had four factories in the city. But all of them were located near the bay, so all of them suffered "total losses." The equipment was swept away together with the facility, and only the building foundation was left.

I could not make sure the safety of our employees because mobile phones were out of service for a while. From the next day after the quake, I was walking 30 minutes every day to the company from my friend's home, where my family had evacuated. Eventually, our employees began to show up at the company one by one, and asked me, "What are your plans from now, boss?" Therefore, on March 24, I called all of our 50 or so employees by our emergency contact network in the parking lot of a hill-side supermarket, and told them, "It was very fortunate that none of us has lost life. I will start the company again from tomorrow."

I think we were probably the earliest in starting restoration activities in Kesennuma. It was a stage far from our normal operation. All of us put on their boots with shovels, and started our work by scraping off the mud piles up in our head office and the factory.

I: When were things getting back to the level that you could work there normally?

O: In April, we fixed our Minato-machi factory and started using it as our temporary facility. As Ishikawadenso was located nearby us, we helped each other such as sharing a generator under no electricity situation.

In the meantime, requests to repair ship machinery damaged by the disaster started coming in. We also undertook various kinds of work requests such as ship inspection and salvage. As we had no tools for those requests, we bought the equipment from an inland DIY center by our light truck. At that time, our activities were more like disaster relief and support rather than a regular work. I don't even remember whether we got paid for that work.

I: Didn't you have a hard time in managing your cash flow?

O: Our local banks and credit unions nearby were all closed, so I went to the branch of a bank in the interiors to withdraw deposits and borrow money. In addition to paying salaries to our employees, we had to pay in cash for things we bought. No employee complained because they knew my situation.

Restoration beyond Individual Reconstruction Efforts

I: In the meantime, I heard that you called shipbuilding-related companies to gather at an early stage, and it led to obtain the Group Subsidy for temporary restoration with extraordinary speed. And it also triggered the birth of MIRAI SHIPS INC., correct?

O: About a week after March 11, I guess, when I went to a supermarket about a 40-minute drive by car toward Ichinoseki City (Iwate Prefecture) to buy things, I happened to see President Yuzo Kidoura of Kidoura Shipyard (at the time). While rejoicing in the fact that we were safe and talking about the damage caused by the disaster to our companies and the industry, we agreed that reconstruction of the shipbuilding-related industries in Kesennuma could not be completed by individual reconstruction efforts, and that we had better talk about it together anyway soon.

By mobile phones which had just began barely connected, we assigned one person to each business line and made them contact with needed people as much as possible. On March 26, in a conference room on the second floor of Kesennuma Pelagic Fishery Association, about 30 companies got together to hold

Shigemasa Onodera

a meeting, which led to the formation of the Kesennuma Reconstruction Committee for Fishing Port Functions.

I: Then the committee took actions to get the subsidy and took a step ahead towards restoration. But I heard that the subsidy did not necessarily reach everyone immediately and simultaneously.

O: We applied for the subsidy at the time of the first supplementary budget in June. But there were nearly 200 groups that had made applications in Miyagi Prefecture, and the competition was severe at a rate of nearly 20 times. However, our application was approved in the first stage, and we got the subsidy. Our group had 58 companies made of not only shipbuilding-related companies but supply companies such as fuel-related, food-related and fishing gear companies, not all of them were granted subsidies simultaneously. It was processed in order, and some companies had to wait until early autumn, I heard.

In fact, such companies as could just barely make it by some repairs were able to get the subsidy even under repair on-going, then be able to accomplish their smooth restoration. However, others who needed to rebuild their new facility had to remove debris first and it took them two to three years, or even four to five years, to complete their restoration. At this stage, however, it was just initiatives toward temporary restoration, and at that time we could only visualize prospects of a temporary restoration at best.

Seven People Who Raised Their Hand

I: After that, how did you move toward the establishment of MIRAI SHIPS INC.?

O: The subsidy grant at this time was only consumed to the restoration work of each. But as long as shipbuilding companies concerned, it was difficult for them to work in their existing sites especially due to land subsidence. A new shipyard was needed to realize their full-scale reconstruction. If they continued their work at the shortened berth, something fraught with danger, they should have been caught in a negative spiral. It might be fine if they intend to quit soon, but it was not an option to those who would continue their shipbuilding business for a long time. From the very beginning, members of the shipbuilding-related industries including me shared the vision of building a shipbuilding complex in a new place and moving there together.

The idea was supported and accelerated by the movement where MLIT was preparing a support for the shipbuilding industry, which in the end took shape in 2013 as "Reconstruction Support Subsidy Project for the Shipbuilding Industry" project. The shipbuilding complex, which all of us hoped, would no longer be a dream if we could get this subsidy. Fortunately, Kesennuma City baked us up by prioritizing our project in their list of the Reconstruction Plan, and we saw hope in securing the construction site.

Since a corporate status was mandatory in applying for the subsidy, the 32 of member companies under, the Kesennuma Shipbuilding and Marine Industries Reconstruction Council, an industry association at that time, assumed making a holding company, under which the participating companies would work individually. It can be compared to a relationship between a large shopping center and specialty stores in there.

In order to realize this, however, a huge amount of money was needed to invest in a shiplift, cranes, factories, and so forth. Even if the subsidy would be accepted, we had to prepare for our own money to finance in addition to the subsidy. At first, we thought of getting a loan from Miyagi Prefecture under tits business upgrade scheme but it required the high interest rate and joint guarantors.

When it came to the stage of establishing the company, I asked all members at the council meeting if there would be someone willing to become joint guarantors. It was the seven presidents including myself who raised their hands, and the seven became the founders of new company in the end. I believe this was the starting point in the establishment of MIRAI SHIPS INC.

Working for Reconstruction

O: At the same time as the start of the establishment of the new company, 18 companies (currently, 23 companies) established the Kesennuma Shipbuilding Complex Cooperative in April 2013 in order to promote the construction of a shipbuilding complex. This Cooperative was responsible for conducting surveys and gathering opinions and views on this project. The seven founders served as directors of the Cooperative, while working on the realization of the plan to establish the new company. The Cooperative was intended to be formed that it would not participate in the new company but would support the new shipyard. Some members of the Cooperative might have factories in the new shipyard while others would only come to work from outside. Certainly, there were some members who were not willing to take another three to five years before the new shipyard got completed. There were such members who had no choice but to leave the Cooperative and rebuild by themselves.

I: Why did you choose to join in making the new company instead of remain in the Cooperative?

O: At first, the plan was not about a merger but about making a holding company under which all of member companies under the Cooperative would share the shipyard. I was not in a position particularly close to shipbuilding companies, but rather in similar position basically as those companies who did not join in the new company.

However, as I had been working to revive the shipbuilding industry since immediately after the quake, it was not my option to leave all of the new company thing to only the shipbuilding companies as their own matter. I felt that I should be responsible for the project, partially for the sake of the revival in Kesennuma. In addition, I did not force other shipbuilding related partner companies to take risks.

I: In your view, what were the thoughts of the top leaders of each shipbuilding company before exploring a path of merger seriously?

O: After the quake it would have been no wonder if any of them decided to go out of their business rather than pass it on to the next generation. Mr. Yoshida of Kosaba Shipyard became ill and was anxious about his health. I guess Mr. Kidoura shared the same view with me because we had worked hard together on an idea of the new company from the beginning but everyone was bound by their own circumstances and there should be some differences in each thought.

I: Specifically, what kind of issues were discussed?

O: When it came to a point where we were heading for, a variety of different opinions came up, such as getting together after five years or continuing as before for another 15 years.

I: What did you think while listening to such discussions?

O: To be honest, I considered that it would be better to go for the merger regardless of any past constraints or employee concerns. But I could not tell any other company to do so from my position. I worryingly listened to the discussion every time and kept thinking that the merger could be a good choice.

I: The members finally chose the merger, not the holding company. Was it because they thought that they could run their business more efficiently?

O: We were running meetings twice a month, and Mr. Yoshikawa and Mr. Sato from Deloitte Tohmatsu joined it by the end of 2013. After their involvement, to my recall, points were made where the merger would eliminate redundancy in the work by the four shipbuilders and since then the momentum for the merger was made.

A Strong Will to Revitalize the Company Together

I: The new factory is about to be completed.

O: Many things happened after we reached an agreement for the merger. It took two years to solve various problems, such as the size of the shipyard, the size of the subsidy to be applied for, and the size of our payments. However, I think, thanks to this tedious process, the presidents of the companies understood that there would be no fall back space and firmly made up their mind. Visits to the officials of MLIT and to various financial institutions with regard to arranging the portion of our payment still remains in my memory. The problem was solved that we were able to get a syndicated loan led by Mizuho Bank, but it was very tough because I was unfamiliar at all with this kind work.

Avoiding any confrontations within our seven members was my top priority at that time. Our competitors are the shipyards across Japan, or in South Korea, Taiwan, or China, which meant that we could not be stranded before entering the ring otherwise we would simply lose.

I: Which do you feel is more, the expectations for the future or the feeling of anxiety about future?

O: Honestly, I don't know exactly what will happen after the four companies get together. You know, people from different companies join and start working together. Some people cannot get along easily with others while some can enjoy being together with others, and some like to achieve something by themselves diligently. It is impossible to have people go through the same things in the same way. What will be needed here is to show strong leadership from management end. Keep sending them a message to us to raise up this new company hands in hands. That will be an important thing to do, sometimes with our heads bowed and not in an enforcing way.

3 | FROM THE FRONTLINE OF SOLVING SOCIAL ISSUES

Support for Reconstruction in Kesennuma

In the wake of the Great East Japan Earthquake, volunteers of the Deloitte Tohmatsu Group gathered and initiated support activities for reconstruction individually in 2011. The Deloitte Tohmatsu Group belongs to a network of the top tier professional firms in the world comprehensively offering audit, tax, legal, and consulting services and, in Japan, has offices nationwide including the three prefectures of Iwate, Miyagi and Fukushima where was affected by the Great Earthquake. Triggered by these voluntary support activities, the group defined that the contribution to the reconstruction and growth of the affected enterprises and economy in the affected areas meets its social mission, and extended its support activities group-wide. In September 2012, the Reconstruction Support Office started.

While the Reconstruction Support Office covers a wide range of activities, the most notable ones are the support to companies in the affected areas for their reconstruction and revitalization by providing its state-of-the art services such as financial analysis and business planning, which are the core of the group, and the contribution to local economies in the affected areas by training local people who are aspired to be the leader in their area and supporting their business ambitions for reconstruction and growth. The support towards the establishment of MIRAI SHIPS INC. in Kesennuma and associating initiatives towards the shipbuilding industry reconstruction can be called as a typical showcase.

It was in August 2012 when Mr. Gentoku Yoshikawa, the head of the Reconstruction Support Office now and one of members at that time, visited Kesennuma. That was his first time visit to the area, accompanied by Mr. Eiji Sato, a member of Sendai Office. A year and a half had already passed after the Great Earthquake. He had heard from local people that the area was getting fairly restored, but he recalls, "I saw shocking scenes in front of me. Honestly, I got impression that reconstruction had yet to be started."

The support on this occasion was to assist in formulating a construction plan for a refrigerated warehouse, a joint-use facility of the Kesennuma Shishiori Fish Processing Cooperative comprising of fish processing companies.[14] Shishiori District is located in the north-central region of Kesennuma City, and has long been known as a cluster area of the fish processing industries. However, it had been devastated by the tsunami following the Great Earthquake, and restoration and reconstruction were urgently required. Mr. Yoshikawa and his team supported the formulation of this construction plan, which led a new refrigerated warehouse completed in August 2015, three years after then. When the support became sort of done, he met the MIRAI SHIPS INC. project, and increasingly devoted himself to it.

Nurturing the Next Generation of Leaders

In order to nurture and produce "leaders with passion, ambition, imagination and action," the Tohoku Future Creation Initiative operated the "Human Resource Development School" together with local governments and chambers of commerce and industry in the affected areas since 2013. Also, in Kesennuma and Minamisanriku District, the "Future Management School" for the development of human resources was started in April of the same year. Mr. Yoshikawa and Mr. Sato became "mentors" of this school to nurture students who would take on leadership roles in the near future. Mr. Takeyoshi Kidoura and Mr. Shigemasa Onodera were the first-generation students of this school, which made Deloitte Tohmatsu involved in the establishment of MIRAI SHIPS INC., as previously described in Chapter One.

Here, I would like to explain in brief about this school. Mentors literally refer to "instructors" or "advisors." They serve as a source of moral and psychological support for students rather than being instructors who just give lectures. Mentors empathically help students to realize the future they envision for themselves. In the limited time of six months, students are required to come up with their own business plan which they really want to see and realize.

---

14 The Kesennuma Shishiori Fish Processing Cooperative was established in August 2012. This organization was established by collaboration among its peer companies in the wake of the Great Earthquake. It started activities, with the support of the government, by raising the affected land at first. After that, it completed a large-scale refrigerated warehouse with a total floor area of 4,235 square meters as a shared facility.

In the first three-month period of the school, students have a lecture-style class every month, where the four subjects of leadership, strategy, marketing, and finance are covered. All of these subjects constitute business essentials that every student should incorporate into his or her own business plan. Students also learn those essentials by way of repeated "mentoring" (regular and continuous exchanges) from their mentors, which may be compared as Zen riddles. The course helps students brush up their concepts and business plans.

Therefore, Mr. Yoshikawa and Mr. Sato, who visited Kesennuma two or three times a month at first, closely communicated with students by email correspondence and telephone conferences. In the latter half of the course, the frequency of their communication got more and more dense, with cutting down on their sleep time, to help students fully prepared for the course-end presentations as part of the graduation ceremony.

The course-end presentation, which marks the culmination of the entire course, is very meaningful to every student in that it is the expression of his or her determination. If the presentation contains any mistake or untruthful content, it may be considered a betrayal to the local residents who may have high expectations to the future activities expressed in the plan. For example, no matter how wonderful and appealing the presentation sounds, it has no value if the presenter thinks that, "In fact, I really don't want to do it that much."

The business plans must be what students work on proactively by themselves, and also must be some solutions to the serious issues of reconstruction after the Great Earthquake and the revitalization of provincial areas. Of course, it goes without saying that the plan is supposed to be both practical and feasible.

During the mentoring period, the mentor repeatedly asks, "Is this business what you really want to do?" "Is this business really necessary for the community?" Students then get "awakened" a lot and start exploring to what extent their thinking will change over the six months and where their real aspirations are. The six month of the school is exactly the process of knowing individual growth.

Staying Close to Thoughts of Two Students

When we look back April 2013, when both Mr. Kidoura and Mr. Onodera enrolled in the school, it was the time that the plan of MIRAI SHIPS INC. as a holding company emerged and the Kesennuma Shipbuilding Complex Cooperative was also organized. At this point, local newspapers had already reported that a subsidy program to support reconstruction of the shipbuilding industry was being prepared by MLIT.

As mentioned in Chapter One, Mr. Onodera had been involved in reconstruction of Kesennuma since immediately after the Great Earthquake. Although he did not own a shipbuilding company, he strongly felt that the revivals of the shipbuilding companies must happen in order for shipbuilding-related industry to ensure its future on the land of Kesennuma. His enrollment to the school coincided with the occasion when he raised his hands to be a core member of a new shipbuilding company. Therefore, it was a natural consequence that Mr. Onodera chose MIRAI SHIPS INC. as the main theme in elaborating his business plan.

Although discussion regarding the merger of the four shipbuilding companies did not exist at all at this point, the reality went to a different direction gradually from what Mr. Onodera had envisioned.

Mr. Yoshikawa commented on this matter saying "Mr. Onodera's business plan was not materialized. It was because the four shipbuilding companies became more and more crucial players in the course of exploring appropriate business plan. Mr. Onodera kept asking to himself how he could contribute to the reconstruction of the shipbuilding industry and where he should position himself to make this happen. I think that he established his own standpoint from these exercises. I would say that the six months at the Future Management School were such time that he strive for defining himself."

Certainly he refined his business plan in his own way during the six-month course because Mr. Onodera had been fully devoted to the MIRAI SHIPS INC. project since before his enrollment to the school. As a result, his plan also let to the framework for the MIRAI SHIPS INC. plan although its final structure ended up differently.

Mr. Onodera mentions, "I decided to do what I could for the industry."

He patiently kept listening to discussions of the four shipbuilding companies which sometimes went on for long hours, and he sometimes served as a negotiator to move things forward. Being a forefront player at times and a supporting player alternately at other times, he advanced the MIRAI SHIPS INC. project.

On the other hand, Mr. Kidoura chose to envision a way to diversify the business of MIRAI SHIPS INC. while leaving the formation plan of MIRAI SHIPS INC. to Mr. Onodera. The theme was regarding development and manufacturing of yachts for the International Yacht Race, the America's Cup. The America's Cup is the world's top yacht race. The Nippon Challenge team from Japan took on the challenge of participating in the race three times in 1992, 1995 and 2000. Recently, the Japanese team led by the Softbank Group at the 35th tournament in 2017 garnered a lot of attention as they had participated in the race after a long interval.

Mr. Kidoura himself belonged to a company developing and manufacturing yachts for the Nippon Challenge for three years from 1998, and experienced building yachts.

"This is my personal dream," says Mr. Kidoura. He believed that reconstruction of the shipbuilding industry in Kesennuma is a must-do task and he needed a very personal dream which would drive him to accomplish the goal with zest and enthusiasm. If you have a dream of building a ship for the America's Cup on your own, you will be able to overcome any challenges in front of you whatsoever and get your business on track faster.

Mr. Kidoura notes, "I'm not the type of a person dictating others by my ideas, but rather want to find such point that others could agree while adjusting my own role and perspective." Mr. Kidoura is the representative director of MIRAI SHIPS INC. and the youngest among the seven business owners. He will definitely continue his role as a coordinator, consistently adjusting and pursue the optimal solution for MIRAI SHIPS INC. by integrating various opinions of all the members.

A coordinator is often forced to relegate one's personal dreams to the back burner, and it may be the case that in the Future Management School, Mr. Kidoura tried to turn his dream into a business plan by the help of Mr. Yoshikawa, not to leave it as just a dream. But it did not happen. All of a sudden, he had to leave from the school for a long time.

Request for Renewed Support

Mr. Kidoura collapsed due to cerebral infarction and had to be admitted in a hospital. This was an unfortunate turn of events but luckily he did not suffer from any major aftereffects. However, he had to be hospitalized from September to mid-February. Right before the period of graduation from the Future Management School, Mr. Yoshikawa and Mr. Sato visited him at the hospital to propose that they would continue to help him until he completed his half-formulated plan. But when they talked to Mr. Kidoura about that, he made an unexpected offer, "Would you help in the reconstruction along with MIRAI SHIPS INC. instead of the America's Cup?" MLIT had already opened applications for the Reconstruction Support Project Subsidy for the Shipbuilding Industry in August. Of course, MLIT was already aware of the plan of the shipbuilding industry in Kesennuma for construction of a new shipyard and was examining the feasibility of the project. In Mr. Kidoura's hospital room, documents and references related to the subsidy from the Ministry lay scattered. Mr. Kidoura urged Mr. Yoshikawa and Mr. Sato to go through them. Mr. Sato mentions, "He said that the size of the project with a subsidy of billions of yen would be huge, something he had never experienced before, so he wanted Deloitte Tohmatsu support the project."

At that point of time, MLIT had already given Mr. Kidoura lots of homework in terms of procedures, and Mr. Kidoura wanted to have some advice from Mr. Yoshikawa, but more importantly wanted to have full analysis on bigger issues such as how the four shipbuilding companies should be organized and what options were available for the integration of the management.

Since then, the phrase, "the four shipbuilding companies" took on special meaning. In April, the seven business owners including Mr. Onodera and Mr. Ishikawa of shipbuilding-related companies as well as Mr. Fujiki of K-Yard, who had relocated to Nemuro in Hokkaido, gathered to establish a new shipyard. However, it was clear that shipbuilding companies had to constitute the core of this huge project and that it and could never be accomplished by a single company. Also, K-Yard, that had already resumed business in Nemuro, could not join the merger immediately. Out of the remaining four companies, two companies, in particular Kidoura Shipyard and Yoshida Shipyard, with capability of handling both new construction and repair of boats, were perceived the leaders in the long-lasting shipbuilding industry in Kesennuma and they were the ones to show their leadership. Moreover, MLIT listed "consolidation of shipbuilding companies and shipbuilding-related companies," namely necessity of business restructuring, as a precondition for provision of the subsidy. Therefore, it was necessary to properly understand what the existing four shipbuilding companies really could do.

Here, a question arises. Why was it Mr. Takeyoshi Kidoura?

The words of Mr. Keigo Yoshida mentioned in Chapter Two provide the clue.

Mr. Yoshida was bewildered when the national government announced introducing subsidy for the reconstruction of shipbuilding. "Many elderly people were working in my company and I think it was more or less the case with other shipyards everywhere. To be honest they might be happier by working for the same company without any changes rather than working in a radically changing environment." Mr. Yoshida eventually decided to participate in the MIRAI SHIPS INC. initiative but it was a very big decision for the man born in 1970 to make the project a real although his ambition to inspire the industry under the new banner of MIRAI SHIPS INC. was genuine.

But on the other hand, Mr. Kidoura was born in 1970 and is younger than Mr. Yoshida by a decade or more. In addition, Mr. Kidoura had in effect already been well trained as senior management with full delegation of business responsibility from the top management even though his title remained senior managing director. Mr. Yoshida must have thought that it would be appropriate to entrust the future of MIRAI SHIPS INC. to this young and energetic leader.

A Business Plan For all Stakeholders

At Mr. Kidoura's request, Deloitte Tohmatsu decided to support the establishment of MIRAI SHIPS INC. through the full efforts of the Reconstruction Support Office. Mr. Yoshikawa and Mr. Sato got involved in the new business plan of MIRAI SHIPS INC. in addition to their role as mentors of the Future Management School.

The first task was to answer Mr. Kidoura's question, "What options are available for business integration?" Since then, Mr. Yoshikawa and his team had more in-depth discussions with five business owners on top of Mr. Kidoura and Mr. Onodera with who they had close relations.

Here, I would like to explain in brief about how Deloitte Tohmatsu participated in the MIRAI SHIPS INC. Project. Mr. Yoshikawa states referring to the role of Deloitte Tohmatsu in the MIRAI SHIPS INC. project "Simply speaking, our role was to provide support by deeply involved in decision-making process and to promote internal and external discussions by way of making highly accountable business plans."

When a company applies for a loan from a bank, they will have to show their business plan and it must naturally be logical and convincing. There must be a proper profit and loss projection without any make-up. The plan will be fully scrutinized by experienced professionals as to whether the business would go profitable and consequently whether repayments will be smoothly made without delay.

It is no exaggeration to say that an answer to Mr. Kidoura's question as to what options were available for managerial integration depended on what kind of business plan could be drawn and whether or not all concerned parties could agree on it.

The business plan was updated every day. All stakeholders, including financial institutions as well as all players involved in the project such as Kesennuma City, MLIT and the banks, paid attention to the validity of the plan.

More specifically, there was a theme of cost reduction in addition to the organization design. Joint purchase could reduce cost in comparison to decentralized purchase, namely buying same material from multiple sources. The same thing goes with reduction of outsourcing costs. Using in-house resources within the four companies as much as possible could save money going outside otherwise.

The challenge that Mr. Yoshikawa and his team took on was to accumulate grounded and vivid numbers which forefront players could agree with, together with the seven business owners. The challenge could be even harder when it comes down to the question whether the seven business owners are the people who could really achieve these numbers, and the question went back to the organizational design such as "What organizational set-up could be required to achieve these numbers?," "What about teamwork?," or "What about the work flow?" Mr. Yoshikawa and his team had started conducting more and more analyses to elaborate the business plan starting the beginning of 2013.

Encouraging Consensus through Detailed Discussions

In December 2013, Deloitte Tohmatsu submitted their conclusion of analysis. "There are two options of company integration, the holding company method and the merger method." According to Mr. Sato, "This was the real start of full-scale discussions regarding business integration and we were allowed to fully participate in the discussion."

After that, the meeting, it was named "the Merger Council," and held twice a month, joined by the seven business owners and both Mr. Yoshikawa and Mr. Sato sometimes via video conferencing. The main topic was to update the business plan, and discussions went on as to which of the two methods, the holding company method or the merger method, would be appropriate in order to ensure the success of the MIRAI SHIPS INC. business.

Mr. Yoshikawa's analysis revealed the point that the most important point of the plan was whether or not the company could have the synergy by integration[15] right after integration happened. That is, if the existing four shipbuilding companies would remain affiliated under the new holding company, it would be the same as not being integrated. For example, if each four company would retain same functionality such as accounting and general affairs separately, then the cost would increase and hurt the integration effect.

In the early stage, they discussed cost reduction under the holding company method, but had a hard time to come up with an effective conclusion, which made the participating members gloomy. Then, Mr. Yoshikawa and his

---

15 Integration synergy refers to an effect in which value generated by integration becomes greater than that when each company operates its business alone.

team presented a concrete cost-cutting plan by intentionally setting a merger option as a viable goal. Consequently, the integration synergy became clear when they chose the merger. Mr. Yoshikawa created a momentum where all members understand the effect of the synergy and could agree with the merger option. Eventually, everyone shared one thought and concluded that the option of a holding company was never viable. It took more than three months to reach this conclusion.

If I describe in this way, you might have an impression that the discussions came to a definitive end successfully at this point, but Mr. Yoshikawa recalls that that was not actually the case.

"I remember that there were swing-backs a couple of times from the worry whether a merger was really a right choice or not. As a merger meant that their respective companies would disappear, they clearly had some hesitation and regrets. The memorandum of understanding was signed in April 2014, but it was just a tentative agreement, which they could easily turn over if they really wanted to. The legally binding merger agreement was signed two years later in March 2016. During that time, I felt like I was in a tense atmosphere and did not know what would happen. Furthermore, I felt really uncertain until April 2018 when the business integration got finally in place."

Essential Human Skills

Mr. Akira Sueishi, who had been seconded from MLIT to Kesennuma City Hall, also participated in the Merger Council from the beginning.

Originally, Kesennuma City intended to use the Reconstruction Grant (Reconstruction Base Maintenance Project for Tsunami Disaster) in order to acquire and develop land for the construction site of the new shipyard. This was not covered by the Reconstruction Support Project Subsidy for the Shipbuilding Industry of MLIT[16]. However, even more than two years after

---

16 The Reconstruction Support Project Subsidy for the Shipbuilding Industry was a subsidy for the reconstruction of a shipyard shared by small and medium shipbuilding-related companies. As it did not cover maintenance or development of the site for construction, Kesennuma City decided to do this through the Reconstruction Base Maintenance Project for Tsunami Disaster that used the Reconstruction Grant from the Reconstruction Agency. (See Chapter One)

the Great East Japan Earthquake, Kesennuma City still had many other problems such as rebuilding housing for local residents and reconstructing the fisheries industry, but the available human skills were limited.

Of course, reconstruction of the shipbuilding industry was a very urgent project as well. In order to make the project a reality, it was necessary to accelerate making plans for the both Reconstruction Grant and the Reconstruction Support Project Subsidy for the Shipbuilding Industry. Furthermore, coordination with local stakeholders was also an urgent requirement. Therefore, Kesennuma City made a request for support to the Maritime Bureau of MLIT. In response to this request, Mr. Sueishi came to Kesennuma City on October 1, 2013.

Mr. Sueishi was sent to Kesennuma City to speed up the reconstruction of the shipbuilding industry. As a city official, Mr. Sueishi worked hard to apply for grants and to acquire land. Meanwhile, he also shared the problems of the seven business owners of MIRAI SHIPS INC., the central figures of project, struggled and began to give them some advice. Mr. Sato mentions, "Mr. Sueishi had started living in Kesennuma since the fall of the previous year, and he got along with local people quite well. He was not much involved in discussions such as talks of a merger or the holding company system, but thanks to his close relationships with members of MIRAI SHIPS INC., he often played intermediate roles by giving appropriate advice from an objective standpoint when the discussions sometimes get emotionally confrontational." While having detailed discussions, Mr. Yoshikawa and his team began to recognize that this support activities for reconstruction were very different from jobs which they had been engaged so far. While they are equally dealing with business plans and analyses with regard to big corporations in the metropolitan areas primarily in Tokyo, the difference was that they were working with local business owners who struggled for reconstruction while facing issues specific to their local areas, They were people who were born in Kesennuma and had lived there, who had their own culture and their own personal relationships, and their personalities were different from any businessmen they had met before. Staying close to them and talking about the future from the same perspective as theirs became a valuable experience with profound lesson with regard to human skills.

Facing the Reality As Is

The memorandum of understanding was signed in April 2014, and MIRAI SHIPS INC. was born on May 1, 2015. However, this was just a framework, and the four shipbuilding companies continued to operate as before at this stage. The Merger Council continued to be held thereafter, and the business plan was repeatedly upgraded.

"I think this time was necessary for the seven business owners to understand the merger process and to fully internalize the concept of this new framework," notes Mr. Yoshikawa.

However, the decision to grant the Reconstruction Support Project Subsidy for the Shipbuilding Industry was taken on May 26 of the same year, and discussions needed to be more specific.

Now with the agreement for the merger of the four companies in place, the four companies were supposed to be absorbed by MIRAI SHIPS INC. and the new company was to operate a new shipyard in the near future. Therefore, stocks must be issued to owners, or shareholders, of MIRAI SHIPS INC. This means that 100 percent of the stocks are allocated to each based on a "shareholder ratio." This ratio is determined by analyzing the financial figures of each company, taking the present, past and to some extent the future of each company into account. Deloitte Tohmatsu conducted a temporary calculation and presented their financial analysis in July 2015. The presidents of the four shipbuilding companies came to know the detailed financial status of other companies for the first time.

They might have chances to look through the financial statements of other companies after they decided to do the business together. According to Mr. Yoshikawa, numbers were inevitably inconsistent due to different accounting standards applied and number processing by tax standards, which were typically the case with small and medium enterprises in general. Deloitte Tohmatsu then conducted their analysis by a consistent accounting standard, and numbers of each company were evaluated by the same measurement. It did happen that what they had dealt as assets were not treated as such. They came to know the actual data of each company for the first time by this analysis.

It seemed that the results were powerful enough to impact everyone. In short, each result of the four shipbuilding companies varied. Sawada Shipyard showed the healthiest financial management, with the other companies following suit. As described in Chapter One, it came from increasing reconstruction demand and the characteristic features of the shipbuilding industry in Kesennuma. The rush of orders from shipowners having the power of setting prices, increased labor cost to meet these orders, and delays in delivery and increased cost due to poor process management skills led to poor profitability.

Having witnessed reality by numbers, Mr. Sawada raised his voice of clearing-up of the debt in each company as the condition for the merger.

Unite Together

Mr. Sawada emphasized that, even though the subsidy would be provided, MIRAI SHIPS INC. itself was going to borrow a lot of money for the new shipyard to start its new business, so unless every company cleaned up their old debt by themselves, they would never see the success.

Deloitte Tohmatsu regarded those debts within a sort of common sense as those debts were intended to finance working capital and capital investment. They were considered to be partially due to temporary problems caused by the Great Earthquake.

In hindsight, Mr. Sawada must have felt a sense of crisis because the financial figures of Kidoura Shipyard and Yoshida Shipyard, both of which were larger than Sawada Shipyard, did not show any better performance than he expected. "Mr. Sawada was very determined. Therefore, the discussion shifted in the middle of its course to the theme how they could pave the way to fully pay back their debts," Mr. Yoshikawa said.

It looked that this was the final task he gave to the four business owners as a mentor, though Mr. Yoshikawa did not say so. It became necessary to find a way to solve this issue together as they were in the same boat. As a result, the four members came to share a sense of solidarity.

"Looking back, I think that a series of moves brought the feelings of the four people together. Mr. Sawada's proposal seemed to have created the major turning point," Mr. Sato recalls.

After the turmoil immediately after the Great Earthquake, the shipbuilding companies generally became somewhat stable. Mr. Kidoura assumed the post of the president by the time of signing the merger agreement. He started full-scale business reforms and started to show results gradually. Also, the short-term business performance of Yoshida Shipyard was getting steady. At the same time, based on the policy of making each company financially healthy, the four companies began to produce the synergy by integration more systematically by way of sharing work and cooperating with each other as much as possible.

The recovery of Kidoura Shipyard was noticeable in particular and Mr. Yoshikawa pointed out that the primary reason can be attributed to the fact that Mr. Kidoura did not conduct major slim-down but rather retained his employees. In the turmoil immediately after the Great Earthquake, Kidoura Shipyard had a sense of mission to protect the fisheries industry, and accepted a large amount of orders for work, and was caught in a difficult situation. At this time, there was an option to fire his employees and outsource it in order to make a profit. However, Mr. Kidoura did not choose it. Not only retaining his employees, he went further and started to increase employment by improving the management skills and order controlling capabilities.

"It was impressive that Mr. Sawada saw this situation and clearly said that employees working at Kidoura were his treasures. He also said that they would be the competitive edge of the new company in the future, and the employees in Kidoura Shipyard were the treasure of Kesennuma," says Mr. Yoshikawa.

Although Mr. Sawada hesitated to agree with the merger till the very end, he was determined to join MIRAI SHIPS INC. Mr. Yoshikawa speculates that the biggest reason for this decision was that Mr. Sawada got willing to reconstruct the shipbuilding industry with young people and that he wanted to nurture those young people by himself. They were determined to cut off all possibilities of retreat after they had confronted each other with complete transparency.

Mirai with Young People

The merger contract was signed in March 2016, and the groundbreaking ceremony for the new shipyard was held in October of that year. The place for the ceremony was packed with people, including the Mayor of Kesennuma, the Deputy State Minister of MLIT and officials from the ministry. Looking at those gathered, Mr. Yoshikawa actually felt that the MIRAI SHIPS INC. project was an important project for Japan. It was simply a large open space now, but he could vaguely imagine a vision of the factory gradually getting into shape with a shiplift, cranes and berths.

"I sometimes get shaken even now by the imagination if someone might suddenly shout that 'I would break this merger discussion!'" says Mr. Yoshikawa with a bitter smile. It is still halfway through. The Merger Council was closed, but instead the board meeting of MIRAI SHIPS INC. is still being held twice a month and Mr. Yoshikawa and Mr. Sato are still participating in it.

What goal will MIRAI SHIPS INC. have achieved when we say the company accomplished its true recovery?

Mr. Yoshikawa says it is when a sort of "soil" of creating employment has been made. Employment is consistently generated in Kesennuma and young people who once left can come back and start to work anytime. Kesennuma City can prosper by creating such an environment that people can make their living with peace of mind. When MIRAI SHIPS INC. becomes the core of the environment, their recovery will be complete. Aging will never stop not only in Kesennuma but also throughout Japan. In order to cope with this emergency situation, Kesennuma must be actively supported by young people, and MIRAI SHIPS INC. must become the central entity in making this happen.

As the first step towards the goal, in April 2018, the four shipbuilding companies had successfully completed Operation Zero Debt and were merged into MIRAI SHIPS INC. They became finally integrated into one company.

Afterword

Gentoku Yoshikawa

Head of Reconstruction Support Office, Deloitte Tohmatsu LLC

Support for Reconstruction in Kesennuma

One of the management philosophies of the Deloitte Tohmatsu Group is "Fairness to Society," which means to ensure fairness of the economy and society and to proactively contribute to its development. If we the Reconstruction Support Office are to rephrases this, it would mean that the role of our group is to contribute to the development of the economy and society by correcting interregional imbalances caused by the Great Earthquake.

Therefore, our support activities for the establishment of MIRAI SHIPS INC. and the construction of the new shipyard are not meant to bring back the status quo of before-the-quake. Reconstruction of the shipbuilding industry will produce a chain of reconstructions of its related companies, fishermen and fish processing companies. Conversely, if the shipbuilding industry does not continue, the industries in Kesennuma themselves could be in danger of ruin. So it is necessary to hasten the process of reconstruction.

The Great East Japan Earthquake has revealed these fundamental problems, but if you see carefully, you will find such a situation in place that the economic base has been gradually damaged due to the declining working population and the aging society with fewer children, even if the disaster would not have occurred. It can be said that the challenges of our modern Japan have already emerged ahead of time. If this situation goes on, a leading fisheries city in Japan may disappear. It is true that this sense of crisis had driven us to act. Therefore, it is important that we should not regard the MIRAI SHIPS INC.'s story as just a matter of the shipbuilding industry in a local area, but as a very symbolic problem in our modern Japan which by chance emerged in an "advanced area" in terms of problems.

Listening and Staying Close became Our Greatest Strength

When I am asked about what Deloitte Tohmatsu has really done in our support activities, visible outcomes are just documents of business plan and minutes of the board meetings and so forth. One may call it trivial, but there is a huge amount of time spent on discussions behind the scene, and I think our communication skills, which allow us to thoroughly discuss until everyone is persuaded, were fully deployed as our strengths.

It may be easier to understand if you use the phrase "local versus Tokyo" as a keyword. Assume that people working in "Tokyo" rushes to save a shipbuilding company of a "local" area. These people are headquartered in Tokyo and many of them work globally. Their way of thinking is dominated by Tokyo logic. But people in a local area have their own local logic. The former, so to speak, can be described as a world of economic rationality, and the latter includes a variety of things, including moral values that are hard to define in words. For example, they are about life, children and grandchildren, a sense of rivalry within the local community, and pride, among other things.

Applying for subsidies, establishing a company, building a new shipyard and so forth. In order to realize these things, the logic of Tokyo would be helpful a lot. However, for example, if we talk down to someone that, "you had better do it because we experts think so," it riles that person. Relying on local logic will not solve problems, but the Tokyo logic is not always correct. Therefore, we thought that "staying close" in an interactive communication was the most important approach when doing this job.

Listen to and stay close. There are different ideas even within the same Japanese. Listen to others' opinion and explore what will be right answer. Think about the meaning of every single word you are speaking to and discover important values in there. It was my pleasure that I went through such experiences and I gained something truly valuable from the work this time.

I think that we need to further sophisticate our communication skills by this experience because it will make us well prepared for the next crisis that may happen someday soon.

Unfortunately, as far as Japan is concerned, we are living in an environment where a big tragedy similar to the Great East Japan Earthquake may occur even tomorrow. Not only we are rightly fearful of such geographical national crises, but we need to be more wise and broad-minded in making all possible preparation to those events even if it occurs today. Suppose that the recovery from the Great East Japan Earthquake will take 10 years, we must set such goal that the recovery from a next great earthquake should be done in five years. I think it is right time for us to make a good preparation for a next hardship.

Become a Company to Lead the Shipbuilding Industry for Fishing Boats in Japan by Making Use of Advanced "Ability to Receive Support"

Let's go back to MIRAI SHIPS INC. I believe that MIRAI SHIPS INC. can be a successful role model of reconstruction after the Great Earthquake. Nothing has been achieved yet, but I think the seven members of MIRAI SHIPS INC. had a very rich experience during the past six years. There would be only a few people who have experienced such a large-scale collaboration with so many supporters. After the long-running six years, they have obtained something they would never get otherwise, by discussing with "outsiders" from Tokyo such as people from MLIT, the Nippon Foundation, and us, Deloitte Tohmatsu, and extending their field of activity.

I think that what the seven business owners have gained through this experience is an enhanced "ability to receive support." If recipients of support do not have ability to receive and connect those supports to success, they will not be able to solve their fundamental problems. This is our insight gained

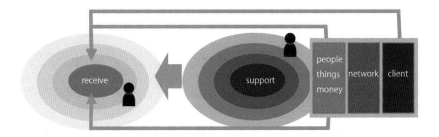

Model for solution of regional issues

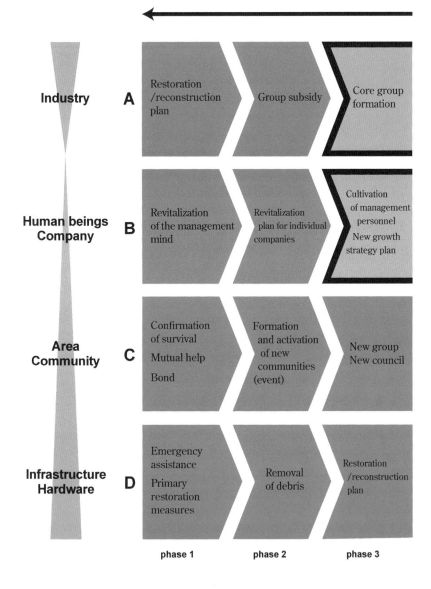

		phase 1	**phase 2**	**phase 3**
Industry	**A**	Restoration /reconstruction plan	Group subsidy	Core group formation
Human beings Company	**B**	Revitalization of the management mind	Revitalization plan for individual companies	Cultivation of management personnel New growth strategy plan
Area Community	**C**	Confirmation of survival Mutual help Bond	Formation and activation of new communities (event)	New group New council
Infrastructure Hardware	**D**	Emergency assistance Primary restoration measures	Removal of debris	Restoration /reconstruction plan

Creation of new value New product New service	New industrial	New industrial innovation New industrial population surpasses existing industrial population	Expansion to other industries
Core company Cultivation (development) Business owner (Business manager) Advancement of human resources	Implementation of the new growth strategy	Realization of the new growth strategy Recovery of profit level (excluding reconstruction demand)	Expansion to companies in other affected areas
New community New group Cooperation and development of the new council	Cooperation with other regions	Establishment of the new economic zone Normalization of domestic and overseas inbound	Expansion to other affected areas
Infrastructure construction (living)	Infrastructure construction (distribution and industry)	New infrastructure innovation Acquisition of innovative technologies and know-how associated with infrastructure construction	Expansion to other affected areas

phase 4   **phase 5**   **phase 6**   **phase 7**

from experiences this time.

Supporters from outside have a wide variety of ways to solve problems. Some supporters may aim to solve the problem by bringing their own resources of "people, things, and money." Other supporters may use their own social network resources. If a supporter is such a company that has a large number of customers, it may build an expanding circle of support by making those customers involved. As described above, supporters try to solve problems by their unique resources, and recipients of support will be able to solve problems more smoothly if they intend to proactively utilize those resources such as "people, things, and money" and/or human networks. As I mentioned above, however, there may be conflicts between the Tokyo logic and the local logic.

Don't you think it would be nice if recipients of support with profound attachment to their place will accept and speak to "outsiders" like us, and think about how they will be able to get support in better way? Consequently they will reach a problem solution much faster by making other surrounding support recipients involved. As times go by, things get better, and outsiders are gone, they may walk by themselves. I think that is exactly the real ability to receive support. The seven business owners of MIRAI SHIPS INC. had acquired this ability.

I think this experience turned to indispensable assets for all of them and will be their great strength towards the future. From now on, they need to collaborate with companies of the Kesennuma Shipbuilding Complex Cooperative, and also need to have various relationships such as competition, harmonization and coordination with shipbuilding companies all over Japan. Of course, it will be important to extend their field by entering into market outside Japan.

The figure shown above is the roadmap for reconstruction to our thought. I think that they have come to around Phases 3 and 4 now, and I hope to see that they will be able to move to Phases 5 and 6 soon by way of getting spotlight to their innovative ship introduction or bringing about new customers by their new technologies.

Another thing I want them to keep in mind is to increase their impact on the local community. Kesennuma is called a "fishery town" today, but I hope it will be someday called a "shipbuilding town." If this happens, I would proudly say that they have finally reached phase 6. Everybody will admit for sure that Kesennuma has fully recovered. I want them to be a leader in the shipbuilding industry for fishing boats in Japan by fully utilizing the ability to receive support that they have acquired from this project. It may or may not happen before I die, but I eagerly want to witness this future to be materialized.

MIRAI SHIPS INC. Timeline

2011	
March	• (11th) The Great East Japan Earthquake occurred.
	• The first meeting of 30 shipbuilding-related companies in Kesennuma was held in a conference room of the Kesennuma Pelagic Fishery Association. After that, it evolved to the "Kesennuma Reconstruction Committee for Fishing Port Functions." They aimed for temporary restoration for repairing and refueling of bonito pole-and-line fishing boats, which would start landing fish in mid-June.
June	• (24th) With cooperation of the Maritime Bureau of the Ministry of Land, Infrastructure, Transport and Tourism, 58 companies of the Kesennuma Reconstruction Committee for Fishing Port Functions applied for a subsidy for the "Group Support Project" of the Small and Medium Enterprise Agency. This became the stepping stone to temporary restoration.
July	• (28th) The organizational meeting of the "Kesennuma Shipbuilding and Marine Industries Reconstruction Council" was held.
	• The "Tohoku Kesennuma Shipbuilding and Shipbuilding-related Industries Business Council" was established. The Nippon Foundation decided to provide a subsidy of 480 million yen to 30 companies of this council in Kesennuma and Minamisanriku-cho as part of the "Revitalization Support Project for the Affected Shipbuilding-Related Companies." (Two large crane trucks and a generator were presented in September of the same year.)
August	• (8th) The Kesennuma Shipbuilding and Marine Industries Reconstruction Council submitted a request titled "Restoration, Reconstruction and Emergence of the Shipbuilding and Marine Industry in Kesennuma" to the Mayor of Kesennuma.
	• Around that time, Mr. Yoshikawa and Mr. Sato of Deloitte Tohmatsu visited Kesennuma for the first time to support reconstruction.
October	• (7th) According to the above request, the reconstruction of the shipbuilding industry became one of the priority projects of the "Kesennuma Reconstruction Plan" in Kesennuma City.
2013	
April	• The "Kesennuma Shipbuilding Complex Cooperative" was established in order to conduct research and gather opinions for construction of a shipbuilding complex and related equipment and to promote its construction. At the same time, discussions on the plan of MIRAI SHIPS INC. as a holding company of seven companies were initiated.
	• The "Future Management School," which Deloitte Tohmatsu participated in as a support activity for reconstruction, was established. Mr. Kidoura and Mr. Onodera participated in it as students.
August	• Applications for the "Reconstruction Support Project Subsidy for the Shipbuilding Industry" started.
September	• Mr. Kidoura was hospitalized (until November).

October	• (1st) Mr. Sueishi was seconded from the Ministry of Land, Infrastructure, Transport and Tourism to Kesennuma City Hall.
November	• The business plan for reconstruction of the shipbuilding Industry in Kesennuma was completed by Mr. Onodera at the Future Management School.
December	• The Merger Council was started. Deloitte Tohmatsu participated in supporting the MIRAI SHIPS INC. project.
2014	
March	• The land construction plan for the construction site for the shipbuilding and repairing facility in Asahi-cho Kesennuma City was approved as the "Reconstruction Base Maintenance Project for Tsunami Disaster."
April	• The four shipbuilding companies concluded the merger agreement.
May	• Around this time, they started requesting financial institutions for funds.
September	• (28th) Some fish processing companies issued an opinion advertisement to request to reconsideration the plan of construction of a new factory in Asahi-cho. It led to a protest campaign against it.
October	• (28th) Kesennuma City established the "Environmental Measures Evaluation Committee" in order to prove environmental safety.
2015	
February	• A syndicated loan was established by eight banks led by Mizuho Bank.
March	• The application for the subsidy was submitted to the Nippon Foundation.
May	• (1st) MIRAI SHIPS INC. was established.
	• (26th) Provision of the Reconstruction Support Project Subsidy for the Shipbuilding Industry was decided.
2016	
March	• The merger contract was concluded.
	• The Nippon Foundation provided subsidy of approximately 800 million yen for the Tsunami-response Shipbuilding Facility Maintenance Project in Kesennuma City.
October	• (21st) The groundbreaking ceremony for the new factory was held.
2018	
April	• MIRAI SHIPS INC. merged with the four shipbuilding companies.
2019	
April	• The new shiplift-type factory was completed.

A SHIP BOUND TO THE FUTURE

Horizon in Social Value Creation
Where Local Insiders Meets Urban Outsiders

2020 年 5 月 25 日　初版発行

著作者	Kazunori JINNAI	ⓒ 2020
発行所	丸善プラネット株式会社	
	〒 101-0051　東京都千代田区神田神保町二丁目 17 番	
	電話　03-3512-8516	
	http://planet.maruzen.co.jp	
発売所	丸善出版株式会社	
	〒 101-0051　東京都千代田区神田神保町二丁目 17 番	
	電話　03-3512-3256	
	https://www.maruzen-publishing.co.jp	
印刷・製本	大日本印刷株式会社	

ISBN 978-4-86345-461-3 C0034